# Illustrated Guide to Houses: Terms, Definitions and Drawings

### Publications

Editor-in-Chief: Cydney L. Capell
Production Director: Marsha E. Tauber

**Copyright © 1995 by Marshall & Swift.
All rights reserved.**

Printed in the United States of America. No part of this book may be reproduced in any form or by any means, or stored in a database or retrieval system, without the prior written permission of the publisher, except in the case of brief quotations embodied in critical articles or reviews. Making copies of any part of this book for any purpose other than your own personal use, is a violation of United States copyright laws.

For information write: Marshall & Swift,
　　　　　　　　　　　P.O. Box 26307
　　　　　　　　　　　Los Angeles, CA 90026-0307

**ISBN: 1-56842-031-5**

# TABLE OF CONTENTS

**INTRODUCTION** .......................... 3

    BATHROOMS .......................... 5

    CEILING FINISHES .................... 9

    DOORS .............................. 13

    ELECTRICAL ......................... 19

    EXTERIOR WALLS .................... 27

    FLOOR FINISHES .................... 35

    HEATING/VENTILATION &
    AIR CONDITIONING ................. 39

    KITCHENS ........................... 45

    PLUMBING .......................... 49

    ROOFING ........................... 53

    SITEWORK .......................... 59

    SPECIALTIES/EQUIPMENT ............ 63

    WALL FINISHES (INTERIOR) ......... 65

    WINDOWS .......................... 69

**USEFUL INFORMATION** ................ 73

    FINANCING ......................... 85

    DEPRECIATION ..................... 91

    GLOSSARY ......................... 99

    INDEX .............................. 115

# GLOSSARY

## RESIDENTIAL CONSTRUCTION NOMENCLATURE

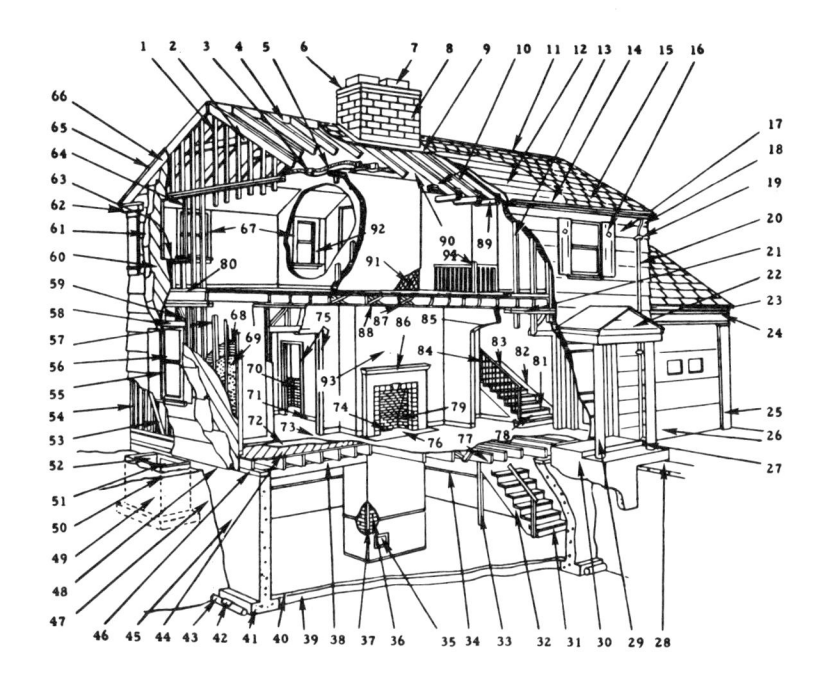

1. Gable stud
2. Collar beam
3. Ceiling joist
4. Ridgeboard
5. Insulation
6. Chimney cap
7. Chimney pot
8. Chimney
9. Chimney flashing
10. Rafters
11. Ridge
12. Roof boards
13. Stud
14. Eave trough or gutter
15. Roofing
16. Blind or shutter
17. Bevel siding
18. Downspout or leader gooseneck
19. Downspout or leader strap
20. Downspout, leader or conductor
21. Double plate
22. Entrance canopy
23. Garage cornice
24. Frieze
25. Doorjamb
26. Garage door
27. Downspout or leader shoe
28. Sidewalk
29. Entrance post
30. Entrance platform
31. Basement stair riser
32. Stair stringer
33. Girder post
34. Chair rail
35. Cleanout door
36. Furring strips

# INTRODUCTION

This book was developed for those readers who need a handy guide to residential construction. Whether for use by new homeowners (or soon-to-be homeowners), insurance professionals, real estate agents, or anyone else with a need for simple, concise information, this book is intended to answer any basic questions you might have. Organized with the user in mind, the book is divided into the various sections of a house. If you want to learn more about an electrical problem, simply turn to that section to learn more about the topic.

In the front of the book is a handy, tear-out guide with over 90 components of a house identified. Each of these terms is thoroughly defined in the glossary at the end of the book. Using this diagram as a guide, if you have questions about a particular section of the house, just turn to that chapter for more information.

A variety of information which you may find useful is organized in the back of the book. Here you can find typical styles of houses identified and described, drawings for determining story heights, and techniques for determining wall and floor measurements and the computations behind those measurements so you can learn these steps yourself. For those readers who use this book while considering the purchase of properties, information on mortgage financing is provided along with mortgage-payment tables. Depreciation tables are available, along with detailed information on life expectancies of homes, to aid you in determining the effective age of a structure.

This book has been developed for you by the researchers at Marshall & Swift. For the past sixty years, Marshall & Swift has been providing building cost data to its customers. From this body of knowledge, the *Illustrated Guide to Houses: Terms, Definitions and Drawings* was developed. We at Marshall & Swift realized that we had provided a variety of detailed materials on both residential and commercial properties for our clients over the years but had not met the needs of those customers who were looking for a simpler, easy-to-use guide. We hope that you will find this book fills that need.

# BATHROOMS

Accessories
Bathtub
Bathtub Enclosure
Bidet
Caulking - Bathtub
Faucet
Medicine Cabinet
Mirror
Shower
Shower Door
Shower Rod
Sink
Toilet
Toilet Seat
Vanity

# BATHROOMS

## TYPICAL BATHROOM LAYOUT

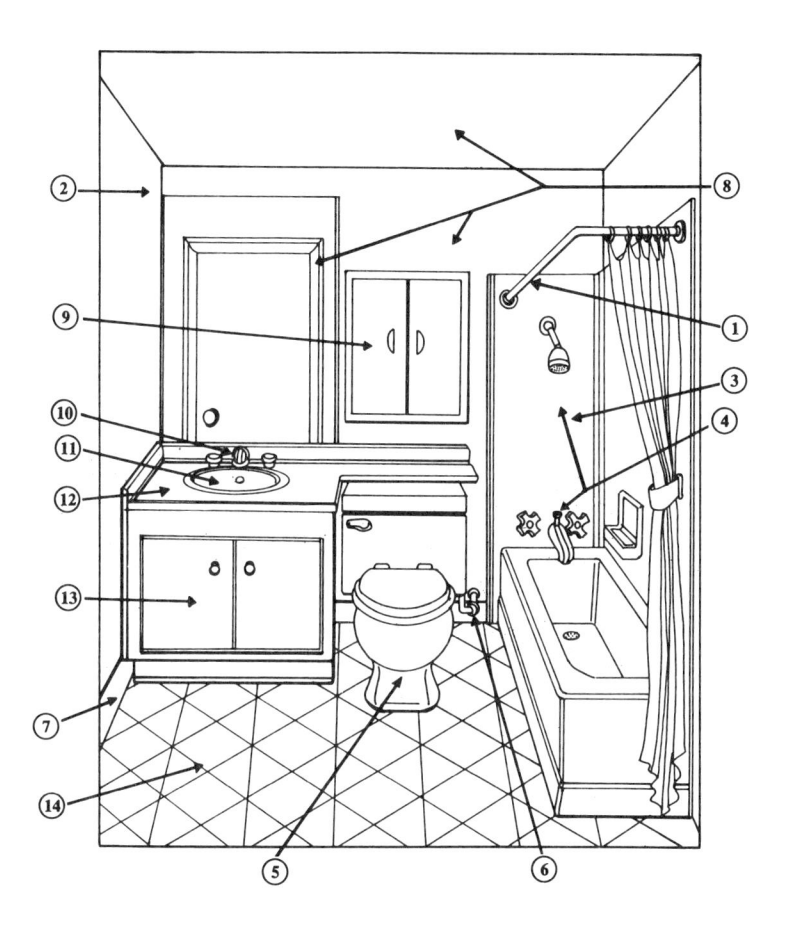

1. Curtain Rod
2. Drywall
3. Fiberglass Tub & Surround
4. Tub & Shower Fittings
5. Tank Type Toilet (2 Piece)
6. Toilet Fittings
7. Baseboard
8. Paint
9. Medicine Cabinet
10. Bathroom Sink Fittings
11. Bathroom Sink
12. Vanity Top
13. Vanity Base Cabinet
14. Vinyl Flooring

# BATHROOMS

## BATHROOM COMPONENT DESCRIPTIONS

**ACCESSORIES SET** – Minor bathroom attachments, such as paper holder, toothbrush holder, soap dish, etc.

**BATHTUB** – A fixed tub used for bathing. The tub can also be part of a modular unit.

**BATHTUB ENCLOSURE** – A metal-framed enclosure made with glass or plastic material.

**BATHTUB/SHOWER COMBINATION** – A one-piece bathtub/shower made of fiberglass.

**BIDET** – A bathroom fixture somewhat resembling a toilet. Used for hygienic washing of the lower private parts of the body.

**CAULKING (BATHTUB)** – A resilient mastic compound between a bathtub and wall or floor surfaces for the purpose of waterproofing.

**FAUCET** – A plumbing valve that combines hot and cold water through one outlet.

**MEDICINE CABINET** – A bathroom storage cabinet for medical supplies, toilet articles, etc.

**MIRROR** – A wall-mounted bathroom mirror.

**SHOWER DOOR** – A shower door made of aluminum-framed glass that operates by sliding on a track or swinging on hinges.

**SHOWERHEAD** – A pipe and nozzle through which water is sprayed.

**SHOWER (OVER TUB)** – Piping, controls and nozzle used to provide water for an over-the-tub shower.

**SHOWER ROD** – A steel tube on which shower curtains are hung.

**SHOWER STALL** – A prefabricated enclosure for a shower.

**SINK (BUILT-IN)** – A built-in bathroom basin.

**SINK (WALL MOUNTED)** – A wall-mounted bathroom basin.

**TOILET SEAT** – The seat portion of a toilet fixture.

**TOILET (FLOOR MOUNTED)** – A toilet fixture whose base is bolted to the bathroom floor.

# BATHROOMS

## BATHROOM COMPONENT DESCRIPTIONS (Cont'd)

**TOILET (WALL MOUNTED)** – A toilet fixture mounted on a wall.

**VANITY (METAL)** – A metal case used primarily to house a bathroom sink, usually with drawers.

**VANITY (WOOD)** – A wood case used primarily to house a bathroom sink, usually with drawers.

# CEILING FINISHES

Acoustical Custom Spray
Acoustical Panel
Acoustical Tile
Furring
Gypsum Board
Insulation
Paint
Plaster
Plastic Panels
Plywood Panels
Stain
Suspension Grid
Wallpaper
Wood Beams
Wood Plank

# CEILING FINISHES

## TYPICAL CEILING SYSTEM FINISHES
## DRYWALL

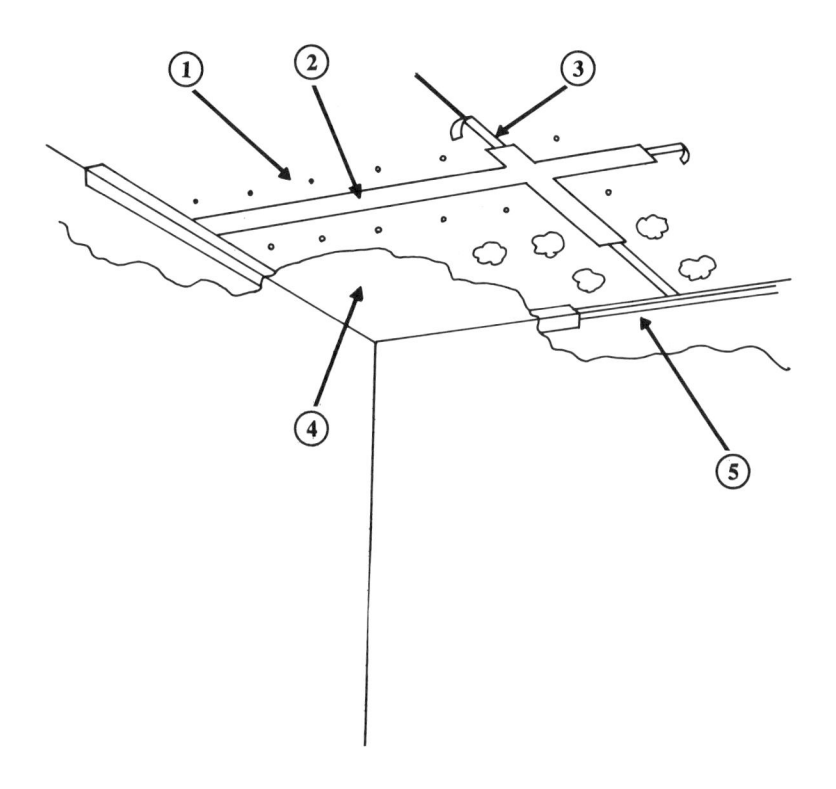

1. Drywall
2. Finish
3. Tape
4. Paint
5. Corners

# CEILING FINISHES

## TYPICAL CEILING SYSTEM
## SUSPENDED ACOUSTICAL

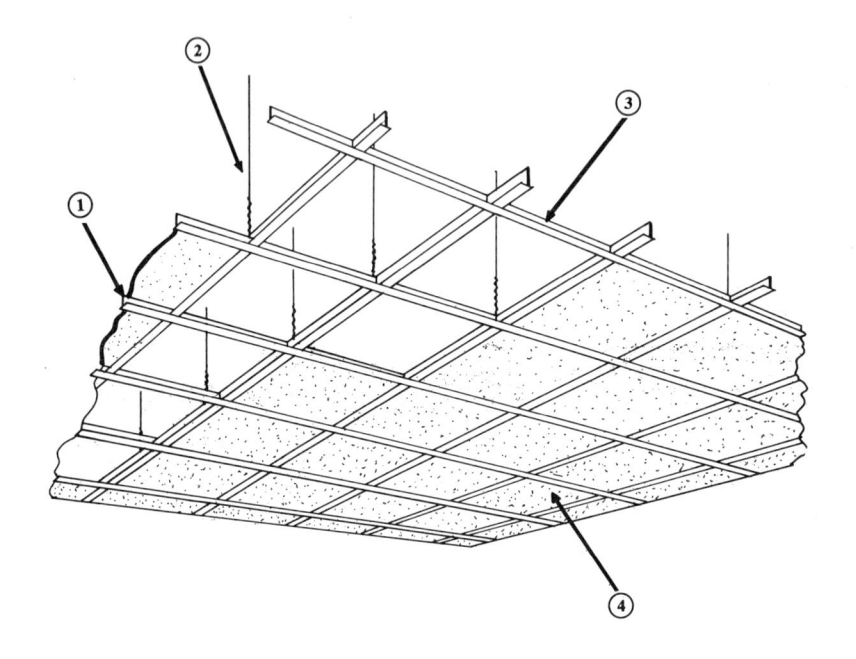

1. Carrier Channels
2. Hangers
3. Suspension Systems
4. Ceiling Board

# CEILING FINISHES

## CEILING COMPONENT DESCRIPTIONS

**ACOUSTICAL CUSTOM SPRAY** – A thin finish coat sprayed onto a ceiling surface. The coat has a granular texture similar to that of acoustical plaster.

**ACOUSTICAL PANEL (ONLY)** – A ceiling cover of a sound-resistant material in panel form.

**ACOUSTICAL PANEL W/SUSPENSION GRID** – A grid system, plus panels, which are made of a sound-resistant material.

**ACOUSTICAL TILE (ONLY)** – A ceiling cover of a sound-resistant material in tile form, usually 12" x 12".

**ACOUSTICAL TILE W/FURRING** – Acoustical tiles applied over furring strips which are fastened to the ceiling.

**FURRING** – Spaced strips of metal or wood which are fastened to the ceiling so that a finished surface can be attached.

**GYPSUM BOARD (STANDARD)** – A common ceiling cover of sheets, typically 4x8, 1/2" to 5/8" thick. It is also known as sheetrock or drywall.

**GYPSUM BOARD (WATER RESISTANT)** – Gypsum board which has been treated to resist the passage of water and moisture.

**INSULATION** – A composite of various types of insulation used above the ceiling. A material with a high resistance to heat flow.

**PAINT** – A colored adhesive coating which can be applied over a variety of surfaces for decoration and protection.

**PAINT (TEXTURED)** – Primer and finish coats of a heavy-bodied paint applied to the ceiling.

**PLASTER (ONLY)** – A fine, gypsum-powdered material that provides a good surface finish.

**PLASTER (ONLY-THINCOAT)** – A ceiling cover of a thin mixture, normally applied over gypsum board to provide a desired surface texture.

**PLASTER W/GYPSUM LATH** – A plaster ceiling cover which is applied with gypsum lath.

# CEILING FINISHES

## CEILING COMPONENT DESCRIPTIONS (Cont'd)

**PLASTER W/METAL LATH** – A plaster ceiling cover which is applied with metal lath.

**PLASTER W/WOOD LATH** – A plaster ceiling cover which is applied with wood lath.

**PLASTIC PANELS (ONLY)** – Panels of plastic used as a ceiling cover.

**PLASTIC PANELS W/SUSPENSION GRID** – Plastic panels installed in a suspended-grid system.

**PLYWOOD PANELS** – A hardwood ceiling cover made of plywood panels.

**REPAIR EXISTING PLASTER** – The repair of hairline cracks, peeling, etc., in a plaster ceiling. Used for small repair areas only.

**STAIN** – A wax-based stain applied as a protective coating on interior wood ceilings.

**SUSPENSION GRID (ONLY)** – A suspension-grid system used for supporting ceiling panels or lighting fixtures.

**WALLPAPER** – A ceiling cover of quality paper.

**WOOD BEAMS** – A nonstructural, wood or synthetic member that horizontally spans the ceiling and is used for ornamentation.

**WOOD PLANK** – A ceiling cover of individual long pieces of wood.

# DOORS

Entry Doors
Frame
French Doors
Garage Doors
Hardware
Metal Doors
Miscellaneous Doors
Paint
Screen Doors
Stain
Storm Doors
Threshold
Trim
Wood Doors

# DOORS

## INTERIOR DOOR SYSTEM

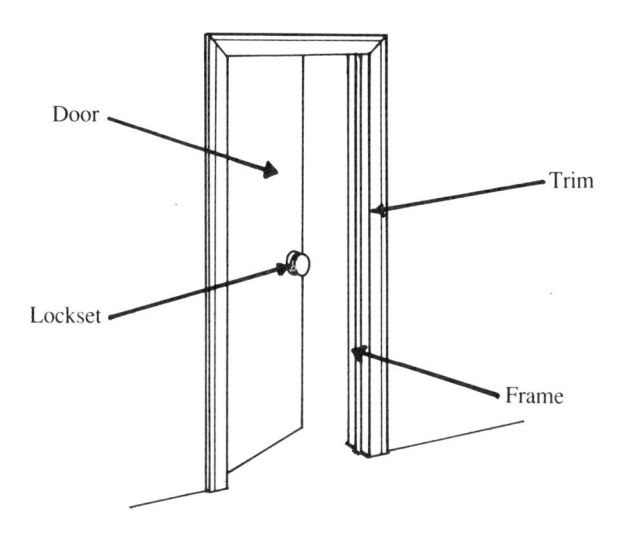

Door

Trim

Lockset

Frame

## EXTERIOR DOOR SYSTEM

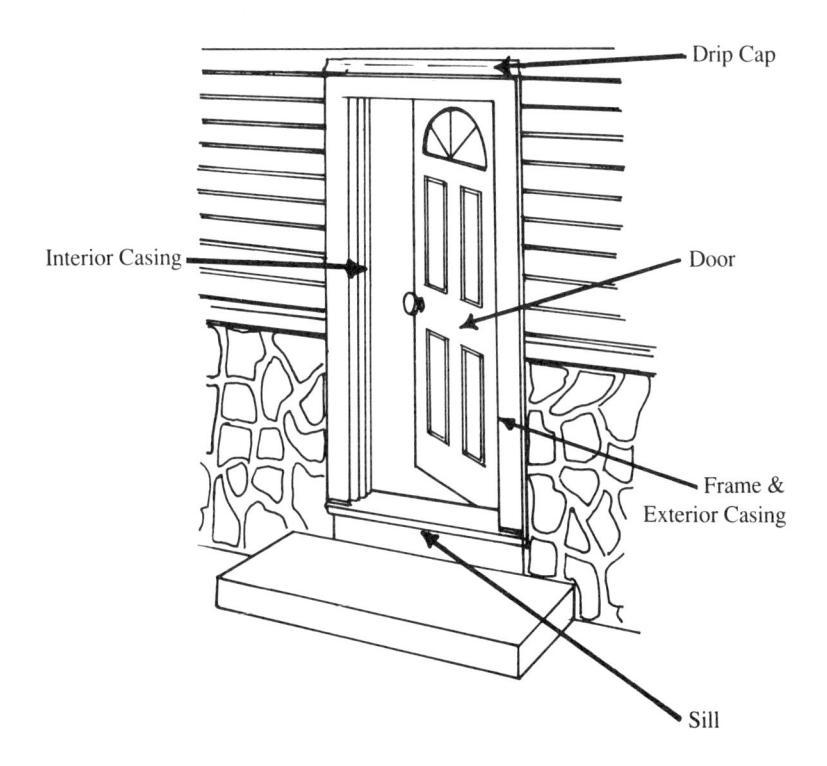

Drip Cap

Interior Casing

Door

Frame & Exterior Casing

Sill

# DOORS

## EXTERIOR DOOR COMPONENT DESCRIPTIONS

**ALUMINUM STORM DOOR** – An auxiliary door placed outside of an existing door.

**DUTCH DOOR** – A door cut horizontally through the lock rail so that the upper or lower part of the door can be opened independently.

**ENTRY DOOR FRAME** – A surrounding door framing unit which includes the jambs, stops, etc. Does not include threshold and trim.

**ENTRY DOOR HARDWARE** – Double-locking hardware with dead bolt and hinges.

**ENTRY DOOR TRIM** – Any visible wood member around the exterior perimeter of the door.

**FRENCH DOOR** – An exterior door having a top rail, bottom rail and stiles with glass panels throughout its entire area.

**GARAGE DOOR (OVERHEAD SINGLE-LEAF)** – A swing-up garage door of single-leaf construction.

**GARAGE DOOR (OVERHEAD SECTIONAL)** – A garage door consisting of more than one section.

**GARAGE DOOR OPENER** – An electric-powered mechanism for opening or closing a garage door.

**METAL STANDARD ENTRY DOOR & HARDWARE** – An exterior metal door including hardware, but excluding the frame.

**METAL STANDARD ENTRY DOOR** – An exterior metal door with an insulating core.

**PAINT DOOR** – A door which has been coated with paint or stain.

**SCREEN DOOR (COMPLETE)** – An auxiliary mesh-cloth wood or aluminum entry door.

**SLIDING DOOR (ALUMINUM/VINYL – (COMPLETE)** – A sliding aluminum or vinyl door with double glass.

**SLIDING DOOR SCREEN (COMPLETE)** – An auxiliary mesh-cloth framed door mounted on a track sliding horizontally.

# DOORS

## EXTERIOR DOOR COMPONENT DESCRIPTIONS (Cont'd)

**SLIDING DOOR (WOOD – COMPLETE)** – A sliding wood door with double glass.

**STAIN DOOR** – A protective stain used on wood doors.

**THRESHOLD** – A strip fastened to the floor beneath the door to cover the floor material joint and to provide weather protection.

**WOOD CUSTOM ENTRY DOOR** – A prefabricated, highly decorative wood door of custom size including frame and hardware.

**WOOD CUSTOM STOCK ENTRY DOOR** – A high-quality prefabricated door in standard sizes, including frame and hardware.

**WOOD STANDARD STOCK ENTRY DOOR** – A smooth-surfaced, solid core, solid hardwood door including frame and hardware.

**WOOD STORM DOOR** – An auxiliary door placed outside of an existing door.

# DOORS

## INTERIOR DOOR COMPONENT DESCRIPTIONS

**CLOSET BIFOLD** – A double-folding closet door.

**CLOSET MIRROR** – A mirrored-surface door.

**CLOSET SLIDING** – A door mounted on a track that slides parallel to the wall.

**DOOR FRAME** – An assembly of two upright members and head over a doorway, enclosing the doorway and providing support on which to hang the doors.

**DOOR HARDWARE** – A complete hardware system including accessories such as knobs, escutcheons, plates, hinges, etc.

**DOOR TRIM** – The interior wood trim of an exterior door or the wood trim on either side of an interior door.

**FLUSH HOLLOW-CORE WOOD DOOR & HARDWARE** – A hollow-core door with hardware but excluding the frame.

**FLUSH HOLLOW-CORE WOOD DOOR (ONLY)** – An interior hollow-core door with either a hardwood or softwood face.

**FLUSH SOLID-CORE WOOD DOOR & HARDWARE** – An interior solid-core door with hardware, but excluding the frame.

**FLUSH SOLID-CORE WOOD DOOR (ONLY)** – An interior door with a solid core usually made of wood.

**FRENCH DOOR & HARDWARE** – A French door including hardware, but excluding the frame.

**FRENCH DOOR (ONLY)** – An interior door having a top rail and stiles, with glass panes throughout.

**PAINT DOOR** – A door which has been finished with paint or stain.

**POCKET DOOR & HARDWARE** – A pocket door with hardware, but excluding the frame.

**POCKET DOOR (ONLY)** – A door that, when opened, slides into a framed wall recess.

# DOORS

## INTERIOR DOOR COMPONENT DESCRIPTIONS (Cont'd)

**RAISED-WOOD PANEL DOOR & HARDWARE** – A raised-wood panel door with hardware, but excluding the frame.

**RAISED-WOOD PANEL DOOR (ONLY)** – A door having stiles, rails and sometimes muntins that form one or more frames around recessed or raised panels.

**STAIN DOOR** – A protective stain used on wood doors.

**THRESHOLD** – A strip fastened to the floor beneath a door to cover the joint where two types of flooring material meet.

# ELECTRICAL

Antenna - TV/Radio
Cable
Ceiling Fan
Conduit
Distribution Subpanel
Doorbell
Door Chime
Fire Alarm
Grounding Rod
Intercom
Lighting Standards
Lightning Arrester
Light Fixtures
Outlet Box
Panelboard
Photocell Device
Receptable
Security Alarm
Service
Switch
Telephone
Thermostat
Timer
Wiring

# ELECTRICAL

## ROOM LIGHTING STANDARDS

**Bathroom**

    Mirrors — Use incandescents or warm white fluorescents on each side of the mirror, about 30 inches apart. Install an incandescent or fluorescent ceiling fixture as well. (For mirrors 36 inches or wider, install three or four incandescents in a 22-inch-minimum width fixture, or install a 36-to 48-inch diffused fluorescent fixture along the top of the mirror.)

    Shower Light — Use an incandescent in a wet-location ceiling fixture.

    Toilet Compartment — Install either a ceiling or wall fixture with an incandescent or fluorescent lamp.

**Bedroom**

    General — Install a ceiling fixture or track lighting of wattage sufficient to provide uniform lighting. Install small ceiling lights in large closets.

    Reading in Bed — Provide an individual incandescent or fluorescent lamp with the bottom of the shade at eye level and 22 inches to the side of the center of the book. As an option, headboard track lighting should provide one incandescent bulb for each person, mounted 30 inches above mattress level.

**Dining Room**

    Chandelier — Provide a total of 300 watts of incandescent lamps. The bottom of the chandelier should be at least 12 inches narrower than the table and 30 inches above the surface.

**Entrance**

    Foyer — In small areas, use incandescent or fluorescent. For larger areas, use incandescent. Consider wall lamps or a chandelier.

# ELECTRICAL

## ROOM LIGHTING STANDARDS (Cont'd)

**Entrance (Cont'd)**

Outside — Flank the door with a pair of incandescent wall fixtures 66 inches above standing level at the door. If only one fixture is possible, mount it at the lock side of the door.

**Hallway**

Ceiling or Wall — Install at least one fixture every 10 feet. Recessed or track accent lighting for wall art is acceptable.

**Kitchen**

Ceiling — Install either incandescent or fluorescent ceiling fixtures sufficient for general lighting.

Sink and Range — Install, over the front edge of the counter, two downlights with reflective flood lamps spaced 18 inches apart. Range hoods require incandescents.

Under Cabinet — Mount as close to the cabinet front as possible. Cover at least two-thirds of the total counter length.

Dinette — Install a pendant incandescent or fluorescent over the table or counter.

**Living & Family**

General — Install a combination of accent and wall-washing track lighting.

Music Stand — Install one reflective or parabolic reflector flood lamp, in a recessed or track fixture, 12 inches to the left and 24 inches in front of the music.

Television — Provide low-level lighting to avoid reflections from the screen.

# ELECTRICAL

## ROOM LIGHTING STANDARDS (Cont'd)

**Living & Family (Cont'd)**

    Game Table — Install one recessed incandescent or fluorescent fixture over each half of the table. For a card or pool table, mount a single-shaded pendant 36 inches above the center of the table.

    Bar — Install recessed or track reflector bulbs, 16 to 24 inches apart, over bars.

**Site**

    Vegetation — Light trees and bushes with spotlights mounted on walls or from ground level. Do not allow light to shine at neighboring houses.

**Stairs** — Provide fixtures at both top and bottom. Control them from each location with three-way switches.

**Study**

    Desk — Position one incandescent or fluorescent lamp with the bottom of the shade 15 inches above the desk and 12 inches from the front edge.

**Track**

    Accent — Ceiling-mounted fixtures should be positioned at a 30-degree angle to prevent light from shining in anyone's eyes. Usually, one fixture is required for each object being accented. To locate the ceiling fixture, the distance from the wall should be 60 percent of the vertical distance from the center of the object to the ceiling.

# ELECTRICAL

## TYPICAL WIRING LAYOUT

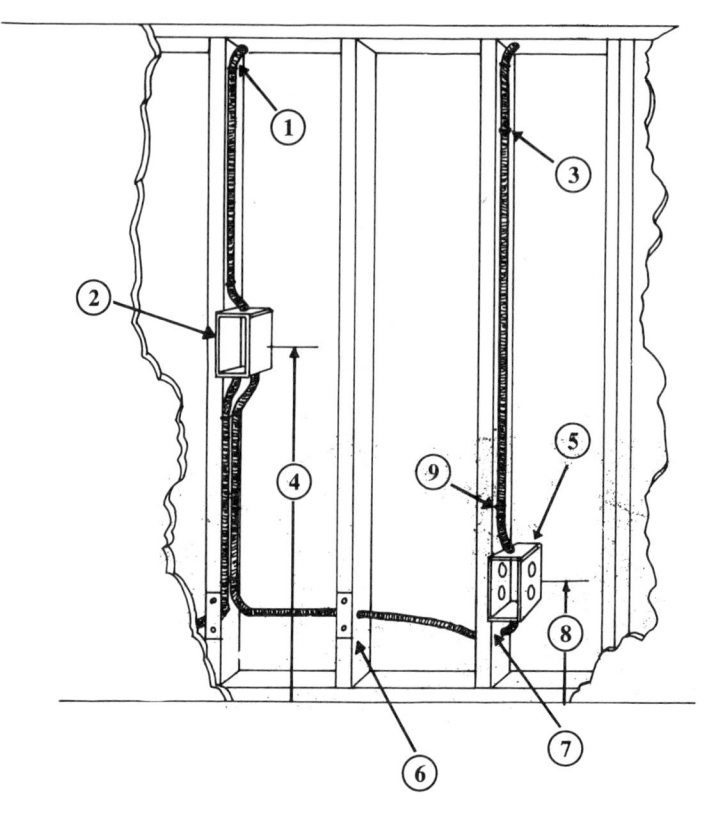

1. Bend Radius
2. Box
3. Staple Cable every 4' 6" minimum
4. Box Height 44" - 48"
5. Metal Box
6. Steel over Cable if within 1 1/2" of Stud Face
7. Hole 1 1/2" from Stud Face
8. Metal Box Height 12" - 18"
9. Staple within 8" of Box

# ELECTRICAL

## ELECTRICAL COMPONENT DESCRIPTIONS

**ANTENNA (TV/RADIO)** – A roof-mounted antenna including the cable and connectors.

**CABLE (WIRE)** – Insulated interior or exterior wires bound together and covered with a nonmetallic sheath. Also known as Romex.

**CANDELABRA** – A fixture with multiple lights suspended from the ceiling, including lamps.

**CEILING FAN** – A fan fixture suspended from the ceiling.

**CEILING FAN W/LIGHT** – A combination fan/light fixture suspended from the ceiling.

**CONDUIT (FLEXIBLE)** – A metal raceway made of an easily bent construction. Includes the fittings, elbows, clamps and wiring.

**CONDUIT (RIGID)** – A raceway of metal pipe of standard thickness that permits the cutting of standard threads or connectors. Includes the fittings, elbows, clamps and wiring.

**DISTRIBUTION SUBPANEL** – An assembly of buses and connections, overcurrent devices, switches and control apparatus, constructed for installation as a complete unit.

**DOORBELL/BUZZER** – A doorbell or buzzer unit, not including the transformer.

**DOOR CHIME** – A chime unit, not including the transformer.

**DOORBELL/CHIME TRANSFORMER** – A small transformer which supplies low-voltage power for operating a doorbell, buzzer or chime.

**EXTERIOR FIXTURE (DECORATIVE)** – An exterior light fixture with decorative features, such as an entrance light.

**EXTERIOR FIXTURE (PLAIN)** – An exterior nondecorative lighting unit.

**EXTERIOR FIXTURE (SECURITY)** – An outdoor, wall-mounted light fixture for security lighting, and usually mounted to prevent vandalism. These may sometimes have a high-intensity bulb.

# ELECTRICAL

## ELECTRICAL COMPONENT DESCRIPTIONS (Cont'd)

**FIRE ALARM CONTROL PANEL** – The base control unit of a fire security system, that can include a built-in local alarm and/or a remote signaling transmitter.

**FIRE ALARM STATION** – A remote signaling device that is hardwired to a control panel.

**FLUORESCENT FIXTURE (4'-LONG STRIP)** – A 4'-long lighting fixture, having either one or two tubes, typically hung from the ceiling.

**FLUORESCENT FIXTURE (4'-LONG SURFACE MOUNT)** – A 4'-long lighting fixture, having either one or two tubes mounted directly to the ceiling.

**FLUORESCENT FIXTURE (DECORATIVE)** – A complete lighting fixture with a louver or diffusing panel, a decorative enclosure and the necessary tubes.

**FLUORESCENT FIXTURE (RECESSED)** – A fluorescent fixture set into the ceiling so that the lower edge of the fixture is flush with the ceiling.

**GROUNDING ROD** – A polarity-type rod used to provide protection to an electrical system.

**INCANDESCENT FIXTURE (DECORATIVE)** – A decorative lighting fixture that uses an incandescent bulb.

**INCANDESCENT FIXTURE (PLAIN)** – An incandescent light fixture, either wall or ceiling surface mounted.

**INCANDESCENT FIXTURE (RECESSED)** – An incandescent fixture set into a ceiling so that the lower edge of the fixture is flush with the ceiling.

**INTERCOM SPEAKER** – A remote speaker of an intercom system that receives from base or remote stations, but cannot transmit.

**INTERCOM STATION** – A remote station of an intercom system that both transmits and receives.

**INTERCOM STATION W/RADIO** – The base station of a radio/intercom system.

# ELECTRICAL

## ELECTRICAL COMPONENT DESCRIPTIONS (Cont'd)

**LIGHTNING ARRESTER** – A roof-mounted device to provide lightning protection.

**OUTLET BOX** – An outlet box with connectors for attaching to the conduit.

**PANELBOARD (CIRCUIT BREAKERS)** – An assembly of buses and connections, overcurrent devices, switches and control apparatus, all of which is constructed as a unit and includes the cabinet.

**PHOTOCELL DEVICE** – A switching device incorporated into an electric circuit that is light controlled.

**RECEPTACLE (110 VOLT)** – A device installed in an outlet box to receive two plugs for the supply of electricity to appliances or equipment.

**RECEPTACLE (220 VOLT)** – A 220-volt contact device installed at the outlet for the connection of a single attachment such as a dryer or range.

**RECEPTACLE (EXTERIOR)** – A plug-in device installed in an outside outlet.

**SECURITY ALARM BASE** – The base control unit of a security system that can include a built-in local alarm and an internal standby battery.

**SECURITY ALARM POINTS** – A remote sensor device to activate an alarm.

**SERVICE (1 PHASE)** – A 200-amp distribution panel with a main switch or circuit breaker.

**SPOTLIGHT (DECORATIVE)** – A decorative light fixture that projects a direct beam of light.

**SWITCH (DIMMER)** – An electrical control device which varies the output of an electrical light fixture.

**SWITCH (EXTERIOR)** – An exterior weatherproof device used to connect and disconnect an electrical circuit.

**SWITCH (3 WAY)** – A wall-mounted device used to open or close a circuit or to change the connection of a circuit.

# ELECTRICAL

## ELECTRICAL COMPONENT DESCRIPTIONS (Cont'd)

**SWITCH (WALL)** – A wall-mounted device used to open or close an electrical circuit.

**TELEPHONE** – A telephone unit with a connection to an outlet.

**TELEPHONE BASE STATION** – The base station for a built-in, in-house telephone system.

**TELEPHONE /TV OUTLET** – An outlet including jack, connections and cover for a telephone or television line.

**THERMOSTAT** – A device activated by temperature changes that controls the furnace and/or air-conditioning output limits.

**THERMOSTAT (PROGRAMMABLE)** – A device which contains a clock system to determine the time periods in which the heating/cooling system controls can be activated.

**TIMER** – A device that manually controls the length of time an electrical circuit will remain on or off.

**TRACK LIGHTING** – A system of lights attached to a section of track and affixed to a wall, ceiling or beam.

**WIRING (110 VOLT)** – Electrical wiring/conductors carrying 110 volts of power.

**WIRING (220 VOLT)** – Electrical wiring/conductors carrying 220 volts of power.

**WIRING COAXIAL (TV/RADIO)** – A coaxial transmission line used in the transmission of television or radio signals.

**WIRING LOW VOLTAGE (DOORBELL/THERMOSTAT)** – A circuit designed for low voltage, such as a doorbell circuit or thermostat.

# EXTERIOR WALLS

Aluminum Siding
Asbestos Siding
Batten Siding Strips
Brick
Caulking
Column - Wood
Concrete
Furring
Hardboard
Insulation
Masonry Block
Paint
Plywood - Textured
Repoint Masonry
Sandblasting
Sanding
Sheathing
Shingles
Stain
Stone Veneer
Stucco
Stud Framing
Trim
Vinyl Siding
Waterproofing
Wood
Wood Shakes
Wood Shingles

# EXTERIOR WALLS

## EXTERIOR WALL TERMINOLOGY
## FRAME SYSTEM

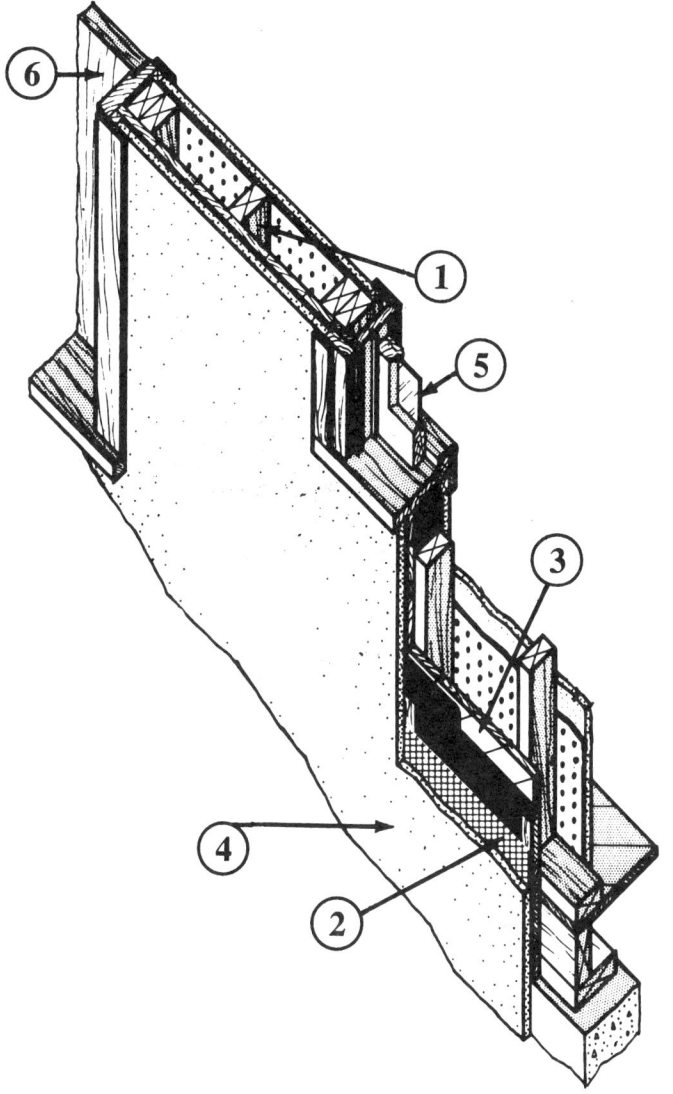

1. Framing (studs)
2. Stucco
3. Sheathing
4. Paint
5. Windows
6. Doors

# EXTERIOR WALLS

## EXTERIOR WALL TERMINOLOGY
## MASONRY SYSTEM

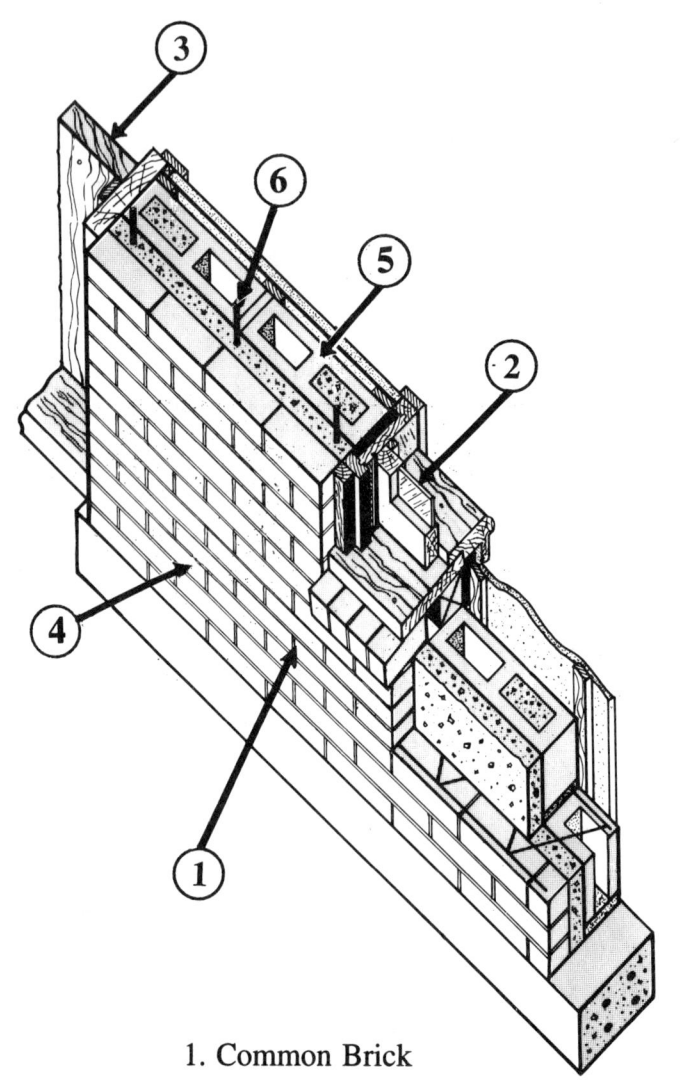

1. Common Brick
2. Windows
3. Doors
4. Waterproofing
5. Block Backup (reinforced)
6. Rebar (reinforcing steel)

# EXTERIOR WALLS

## BRICK WALL PATTERNS

Running

English

Common

Dutch

Common With Flemish Headers

Flemish Cross

Garden Wall

Flemish

# EXTERIOR WALLS

## STONE WALL PATTERNS

### LOCAL STONE

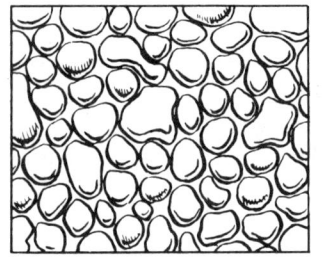

Cobble

Rubble

### ASHLAR FACING

Coursed Saw Bed

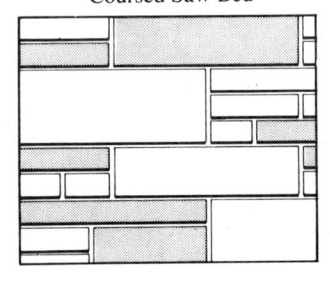

Random Rough Bed

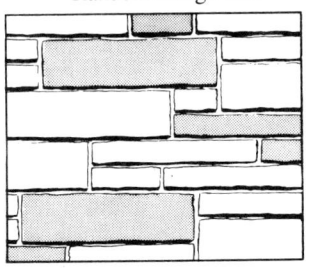

# EXTERIOR WALLS

## TYPICAL EXTERIOR WALL SYSTEMS
## BRICK OR STONE VENEER

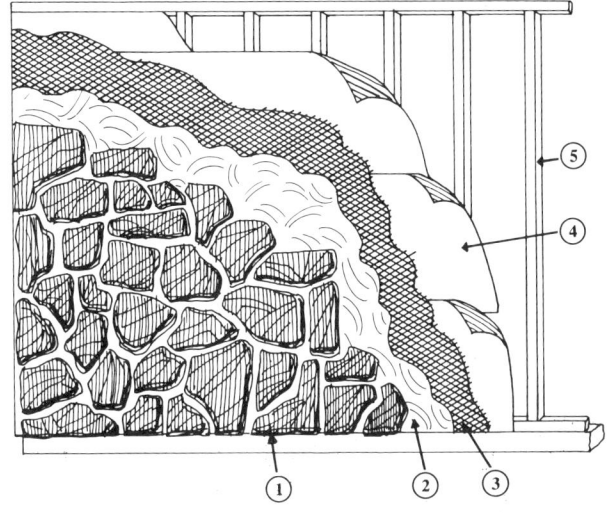

1. Brick or Stone
2. Scratch Coat
3. Metal Lath
4. Building Paper
5. Stud

## ALUMINUM OR VINYL SIDING

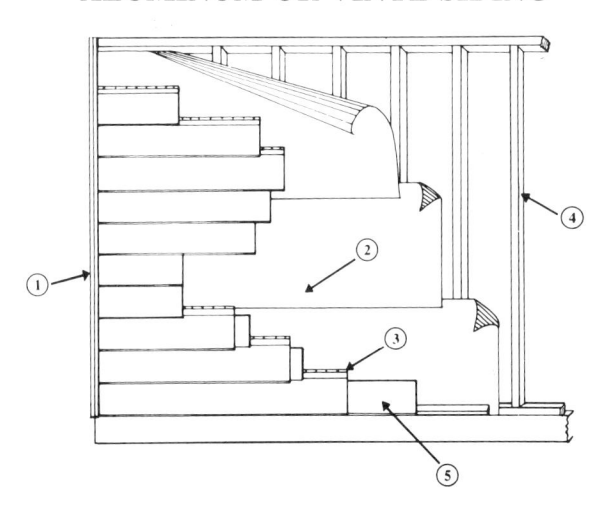

1. Trim
2. Building Paper
3. Aluminum or Vinyl Horizontal Siding
4. Stud
5. Backer, Insulation Board

# EXTERIOR WALLS

## EXTERIOR WALL COMPONENT DESCRIPTIONS

**ALUMINUM SIDING** – An exterior wall covering typically composed of 1" x 8" aluminum siding applied over a stud wall.

**ASBESTOS SIDING** – An exterior wall covering composed of asbestos siding applied over a stud wall.

**BATTEN SIDING STRIPS** – 1" x 2" wood strips nailed vertically to siding sheets at the butted joints.

**BRICK (SOLID)** – An exterior wall consisting of concrete or clay bricks.

**BRICK VENEER** – A nonbearing outside wall facing of brick providing a decorative surface.

**CAULKING** – Installation of a resilient mastic compound used to seal cracks, fill joints, prevent leakage and/or provide waterproofing.

**COLUMNS (WOOD)** – An architectural, hollow or solid wood column that serves as ornamentation. May or may not be load-bearing.

**CONCRETE** – An exterior wall of reinforced concrete.

**FURRING** – Wood strip spacers that are fastened to a wall to provide a flat plane upon which siding or other surface material may be installed.

**HARDBOARD BOARDS** – An exterior wall covering typically composed of 1" x 12" hardboard applied over a stud wall.

**HARDBOARD PANELS** – An exterior wall covering typically composed of 4' x 8' hardboard sheets applied over a stud wall.

**INSULATION (BATT)** – A flexible blanket or roll-type insulation installed between studs in frame construction.

**INSULATION (BLOWN-IN)** – A cellulose insulation material blown-in between the wall spaces of a frame construction.

**INSULATION (RIGID)** – A structural building board applied to walls to resist heat transmission.

**MASONRY BLOCK** – An exterior wall consisting of concrete masonry units.

**PAINT** – A colored, adhesive coating which can be applied over a variety of surfaces for decoration and protection.

# EXTERIOR WALLS

## EXTERIOR WALL COMPONENT DESCRIPTIONS (Cont'd)

**PLYWOOD (TEXTURED)** – An exterior wall covering of textured plywood (T1–11) panels applied over a stud wall.

**REPOINT MASONRY** – The removal and replacement of mortar from between the joints of masonry units.

**SANDBLASTING** – The use of sand, propelled by an air-blast unit, to remove dirt, rust, paint, or to decorate the surface with a semirough texture.

**SANDING** – The removal of damaged wall finish by sanding.

**SHEATHING** – Plywood sheets attached to stud framing to provide backing to exterior wall materials and add to frame rigidity.

**SHINGLES (MISCELLANEOUS)** – An exterior wall comprised of pieces of any number of materials, such as wood, fiberglass, cement asbestos, etc., applied over a stud wall.

**STAIN** – A protective coating applied to exterior wood.

**STONE VENEER (IMITATION)** – A nonbearing outside wall facing of synthetic stone, providing a decorative surface.

**STONE VENEER (NATURAL)** – A nonbearing outside wall facing of thin natural stone, providing a decorative surface.

**STUCCO (ON FRAMING)** – An exterior wall covering of stucco applied to a stud wall, including lath or wire or plaster.

**STUCCO (ON MASONRY)** – An exterior wall covering of stucco applied to a masonry surface.

**STUD FRAMING** – An exterior wall constructed using wood studs, plates, firestops, and bracing.

**TRIM (WOOD/METAL)** – Any visible finishing component, usually of metal or wood (cornices, fascias, etc.).

**VINYL SIDING** – An exterior wall covering of typically 1" x 8" extruded vinyl applied over a stud wall.

**WATERPROOFING (BUILDING PAPER)** – A water-impervious paper, such as tarpaper, applied to a wall to prevent the passage of moisture.

**WATERPROOFING (CEMENT PARGING)** – A coat of portland cement mortar applied to the earth side of a foundation or basement walls to provide dampproofing on masonry facing.

# EXTERIOR WALLS

## EXTERIOR WALL COMPONENT DESCRIPTIONS (Cont'd)

**WATERPROOFING (HOT MOPPED)** – The use of one or more hot-applied coatings or layers of a material to an exterior wall to prevent the passage of moisture.

**WATERPROOFING (PLASTIC SHEETING)** – The use of a plastic sheet applied to a wall to prevent the passage of moisture.

**WOOD SHAKES** – An exterior wall covering comprised of any thick, hand-split shingle or clapboard, usually edge-grained, applied over a stud wall.

**WOOD SHINGLES** – An exterior wall covering of wood shingles laid at either 7 1/2" or 11 1/2" exposure to weather.

**WOOD SIDING (BEVEL)** – An exterior wall covering typically consisting of 1/2" x 8" wood boards whose beveled cross sections enable each board to overlap one another.

**WOOD SIDING (CLAPBOARD)** – An exterior wall covering comprised typically of 1" x 8" wood boards applied over a stud wall.

# FLOOR FINISHES

Brick Pavers
Carpet
Carpet Pad
Floor Sleepers
Linoleum
Marble
Resilient
Sand and Finish Floor
Slate
Tile - Asphalt
Tile - Ceramic
Tile - Quarry
Tile - Rubber
Tile - Vinyl
Underlayment - Hardboard
Wood

# FLOOR FINISHES

## TYPICAL FLOORING SYSTEM

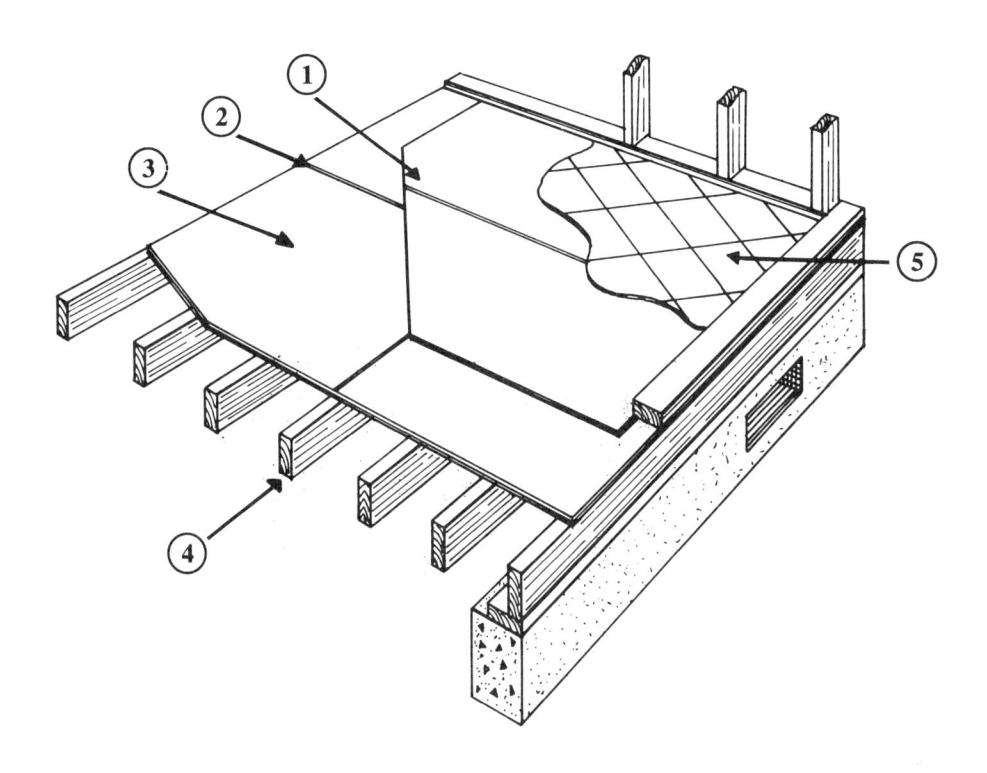

1. Underlayment Panels (if required)
2. Provide 1/32" Space Between Panel Edges
3. Subfloor
4. Floor Joists
5. Flooring Material

# FLOOR FINISHES

## FLOOR COMPONENT DESCRIPTIONS

**ASPHALT TILE** – A floor surfacing unit composed of asbestos fibers, mineral fillers and pigments.

**BRICK PAVERS** – A floor surface of rectangular-shaped blocks of special fired clay.

**CARPET (INDOOR/OUTDOOR)** – Carpeting constructed from synthetic material. Designed to be used either inside or outside a building.

**CARPET (STAIR)** – The material and installation of carpeting on stairs.

**CARPET (SYNTHETIC)** – Carpeting constructed of synthetic materials such as polypropylene, nylon and acrylic.

**CARPET (WOOL)** – Carpeting constructed of animal fiber.

**CARPET PAD** – Rolls of urethane or similar material, used as cushioning under carpet.

**CERAMIC TILE** – A floor surface unit whose body is made of vitrified clay, either mud set or mastic set.

**FLOOR SHEATHING (BOARDS)** – Wood boards which provide a base for the application of a floor surface.

**FLOOR SHEATHING (PLYWOOD)** – Structural plywood which provides a base for the application of a floor surface.

**FLOOR SLEEPERS** – Horizontal wood members that are laid on a concrete slab and to which the flooring is attached.

**LINOLEUM** – A resilient floor surfacing material manufactured in large sheets. Thickness ranges from .125" to .220".

**MARBLE** – A floor surfacing constructed of natural marble.

**QUARRY TILE** – A floor surface manufactured from natural clay or shales, usually unglazed. The tiles are normally 6" or more in surface area and approximately 1/2" to 3/4" thick.

**REGROUT TILE FLOOR** – Reapplying cement mortar between tile joints on a floor surface.

# FLOOR FINISHES

## FLOOR COMPONENT DESCRIPTIONS (Cont'd)

**RESILIENT** – A floor covering classification which includes, but is not limited to, asphalt, cork, rubber, and vinyl.

**RUBBER TILE** – A floor surfacing unit composed of rubber.

**SAND & FINISH DAMAGED FLOOR** – Removal of old floor finish through sanding, and the application of a new finish.

**SAND & FINISH NEW FLOOR** – Sanding and applying the initial finish to a newly laid wood floor.

**SLATE** – A floor surfacing made of thin tiles of slate, usually mud set or mastic set.

**UNDERLAYMENT (HARDBOARD)** – Hardboard applied over the existing floor cover allowing installation of a new floor surface.

**VINYL TILE** – A floor surface unit, usually 12" x 12", composed principally of polyvinylchloride set in mastic.

**WOOD (HARDWOOD)** – A floor covering made of hardwoods, such as oak, beech, birch, pecan, etc.

**WOOD (SOFTWOOD)** – A floor covering made of softwoods, such as pine, fir, etc.

**WOOD PARQUET TILE** – A floor covering of inlaid hardwood (tongued, grooved and end-matched), of short lengths or individual pieces, and usually set in geometric patterns.

# HEATING, VENTILATION & AIR CONDITIONING

Air Conditioner
Air Duct
Air Exchanger
Air Intake Grille
Air Purifier
Air Register
Blower - Ventilation
Boiler
Chimney - Metal
Dehumidifier
Exhaust Fans
Expansion Tanks
Fan - Window
Furnace
Heat Pump System
Heaters
Humidifier
Oil Tank
Package Unit
Radiant Ceiling Heat
Radiant Floor Heat
Radiator - Hot Water
Vent - Dryer
Vent Stack
Ventilator - Attic

# HEATING/VENTILATION & AIR CONDITIONING (HVAC)

## HEATING & AIR-CONDITIONING SYSTEMS
## FORCED-AIR FURNACE

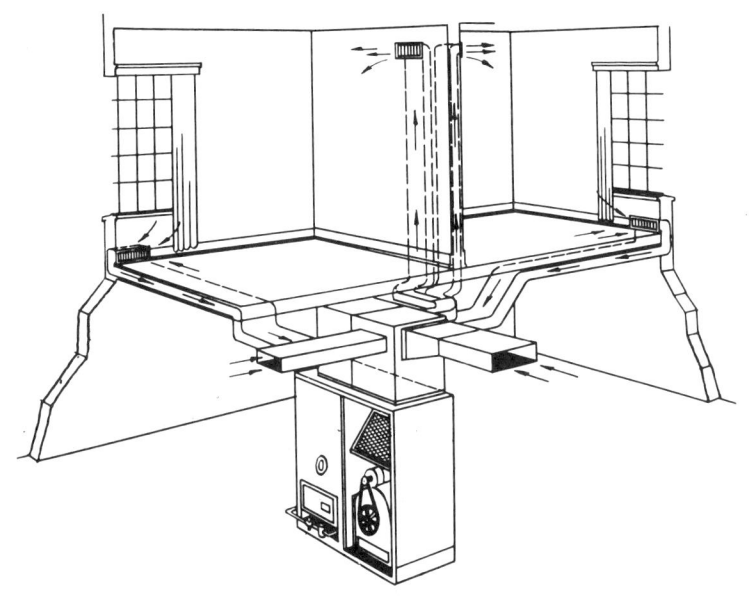

## HEAT PUMP

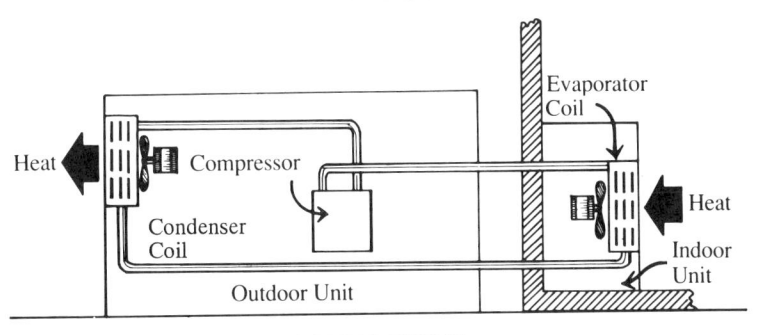

COOLING CYCLE

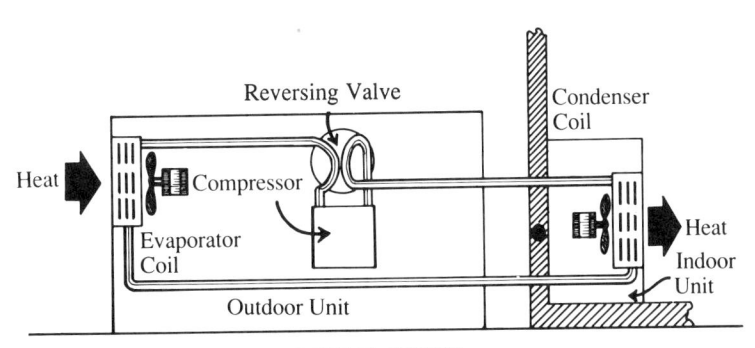

HEATING CYCLE

# HEATING/VENTILATION & AIR CONDITIONING (HVAC)

## HEATING & AIR-CONDITIONING SYSTEMS
## HOT WATER

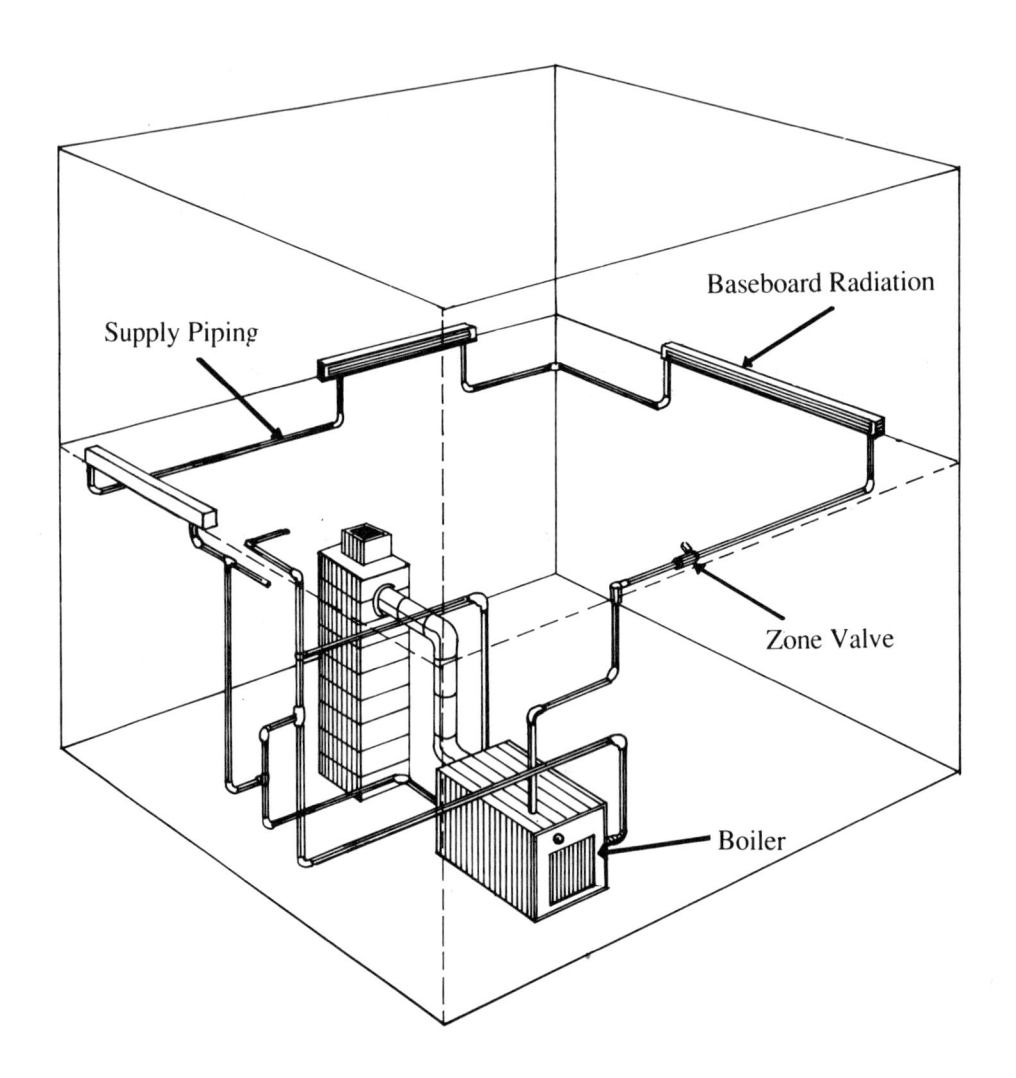

- Baseboard Radiation
- Supply Piping
- Zone Valve
- Boiler

# HEATING/VENTILATION & AIR CONDITIONING (HVAC)

## HEATING & AIR-CONDITIONING SYSTEMS

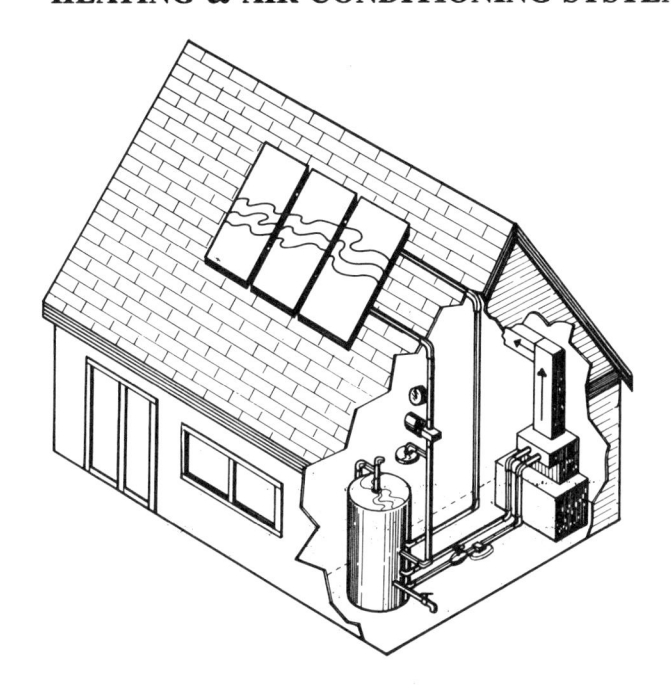

### SOLAR HEATING SYSTEMS

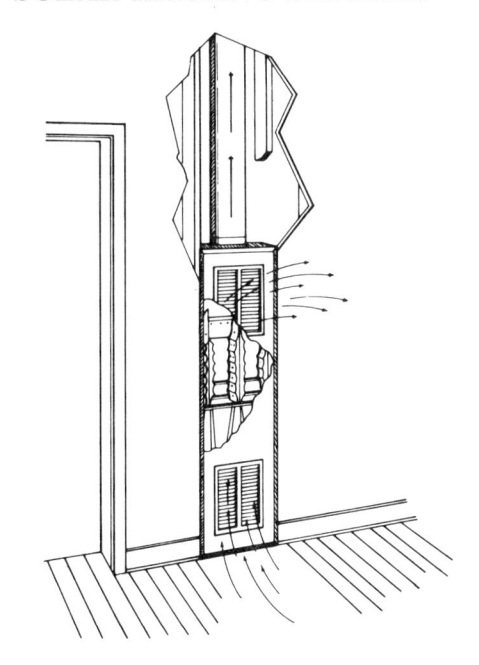

### WALL FURNACE

# HEATING/VENTILATION & AIR CONDITIONING (HVAC)

## HVAC COMPONENT DESCRIPTIONS

**AIR CONDITIONER (WINDOW)** – A unit designed to be installed in a window opening or wall.

**AIR DUCT** – A duct, usually fabricated of sheet metal, fiberglass, or vinyl, sometimes wrapped in fiberglass for insulation.

**AIR EXCHANGER** – A device located in air ducting. Used in conjunction with an HVAC system for transferring temperature from one air flow to another.

**AIR INTAKE GRILLE** – The frame covering the opening of an air intake duct.

**AIR PURIFIER (ELECTRONIC)** – An electronic device for cleansing air. Used in combination with an HVAC system.

**AIR PURIFIER (FILTERED)** – A filter device for cleansing air. Used in conjunction with an HVAC system.

**AIR REGISTER (SUPPLY)** – A grille having a damper for the regulation of the air supply to a room from a heating/cooling unit.

**AIR REGISTER (RETURN)** – A grille covering a vent from a room into a return air duct leading to a heating/cooling unit.

**BLOWER (VENTILATION)** – A fan used for the movement of air through a duct system. Used in conjunction with HVAC systems.

**BOILER (HOT WATER)** – A central unit in which hot water is heated by gas, oil or electric and circulated by piping through the building.

**CHIMNEY (METAL)** – A preconstructed metal flue used to vent a fireplace or furnace.

**DEHUMIDIFIER** – A device for removing moisture from the air. Used in conjunction with an HVAC system.

**EXHAUST FAN (ATTIC)** – A fan used to withdraw and discharge hot air from an attic.

# HEATING/VENTILATION & AIR CONDITIONING (HVAC)

## HVAC COMPONENT DESCRIPTIONS (Cont'd)

**EXHAUST FAN W/SHUTTER (ATTIC)** – An attic exhaust fan with an adjustable cover.

**EXHAUST FAN (KITCHEN/BATHROOM)** – A device to remove air from a kitchen or bathroom.

**EXHAUST FAN (WHOLE HOUSE)** – An attic fan designed to circulate air throughout a structure.

**EXPANSION TANK (HOT WATER)** – A reservoir that holds the water overflow from heat expansion.

**EXPANSION TANK INSULATION** – The insulating material encasing a hot water expansion tank.

**FAN (WINDOW)** – A fan set into a window opening.

**FURNACE (FORCED AIR)** – A warm-air central heating system equipped with a blower to circulate the air. Air can be heated by gas, oil or electric.

**HEAT PUMP SYSTEM** – An electrical refrigeration system that provides reverse cycle operations for cooling and heating.

**HEATER (ELECTRIC BASEBOARD)** – An electrical heating system in which heating elements are installed in panels along the base of the wall.

**HEATER (HOT WATER BASEBOARD)** – A heating system in which centrally heated water is circulated through panels along the base of a wall.

**HEATER (WALL)** – A self-contained hot air heater permanently attached to a wall. Air can be heated by either gas or electric, sometimes fan forced.

**HUMIDIFIER** – A device used to add moisture to the air and used in combination with an HVAC system.

**OIL TANK** – A fuel oil storage container for a residential structure. Costs include the tank only.

**OIL TANK SUPPLY LINE** – A pipe that supplies fuel oil to a furnace from a central supply tank.

# HEATING/VENTILATION & AIR CONDITIONING (HVAC)

## HVAC COMPONENT DESCRIPTIONS (Cont'd)

**PACKAGE UNIT** – A combination heating/cooling package. Costs include the compressor but exclude the controls, ducting and piping.

**PIPE (HOT WATER BRANCH)** – Distribution pipes of a hot water system that carry and return hot water throughout the system.

**PIPE (HOT WATER MAIN)** – A large or main pipe that supplies a number of distribution pipes with hot water or return water from a central heating plant.

**RADIANT CEILING HEAT** – A heating system using an electric grid in the ceiling plaster.

**RADIANT FLOOR HEAT** – A system of heating whereby hot water flows through piping installed within a floor.

**RADIATOR** – A unit that radiates heat by the use of hot water. The unit is usually exposed to view in the room.

**RADIATOR (FIN TUBE)** – A heating system in which water that is centrally heated is circulated through finned tubes in an enclosure along a wall.

**THERMOSTAT** – A device activated by temperature changes that controls the furnace and/or air-conditioning output limits.

**THERMOSTAT (PROGRAMMABLE)** – A device which contains a clock system to determine the time periods in which heating/cooling controls are to be activated.

**VENT (DRYER)** – Ducting between a dryer and a vent.

**VENT STACK** – A vertical vent pipe installed through the roof, primarily for the purpose of conducting and discharging air.

**VENTILATOR (ATTIC)** – A mechanical fan, located in the attic space of a residence, that is used to vent attic air.

# KITCHENS

Base Cabinet
Blender - Food Center
Countertop
Dishwasher
Faucet
Freezer
Microwave Oven
Oven
Range
Range Hood
Refrigerator
Sink
Trash Compactor
Wall Cabinet

# KITCHENS

## TYPICAL KITCHEN LAYOUT

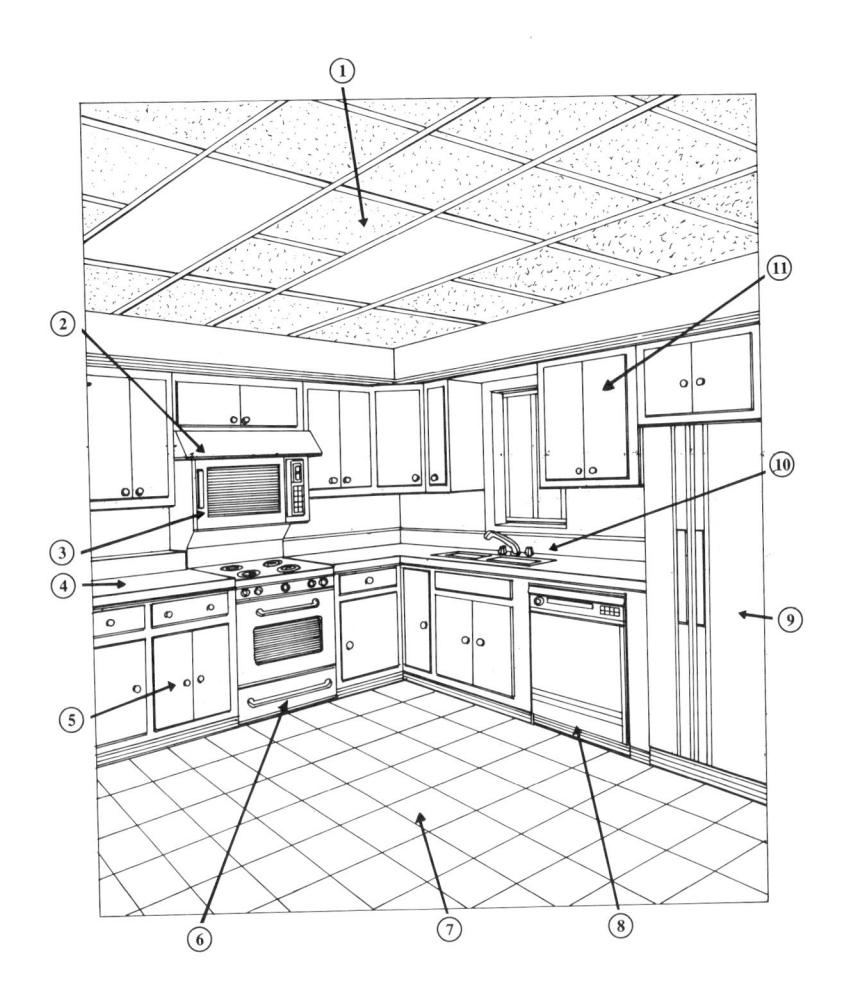

1. Drop Ceiling
2. Range Hood
3. Microwave Oven
4. Formica Countertop & Backsplash (or tile counter)
5. Base Cabinets
6. Cooking Range/Oven
7. Vinyl Floor Tile
8. Dishwasher
9. Refrigerator
10. Sink w/Faucet
11. Wall Cabinets

# KITCHENS

## KITCHEN COMPONENT DESCRIPTIONS

**BASE CABINET** – A kitchen storage case usually made out of wood or metal.

**BLENDER (FOOD CENTER)** – A food-preparation device and control unit built into a kitchen counter.

**COUNTERTOP** – A top or working surface of a kitchen base cabinet made out of a variety of materials.

**DISHWASHER** – A self-supporting appliance for cleaning dishware.

**DISHWASHER (BUILT-IN)** – An appliance for cleaning dishware that is built into a kitchen cabinet.

**FAUCET** – A water outlet control device.

**FAUCET (COMBINATION)** – A plumbing valve that combines hot and cold water through one outlet.

**FREEZER** – A self-supporting freezer.

**GARBAGE DISPOSAL** – An electric device for grinding waste food prior to its entering the sewer pipe.

**MICROWAVE OVEN** – A self-supporting microwave oven.

**OVEN** – A single or double oven that is built into a kitchen cabinet.

**OVEN (MICROWAVE BUILT-IN)** – A microwave oven that is built into a kitchen cabinet.

**RANGE (COOKTOP)** – A burner unit that is recessed into the surface of a kitchen counter.

**RANGE HOOD** – An exhaust hood over a kitchen stove or range.

**RANGE/MICROWAVE COMBINATION** – A self-supporting stove and microwave attached as one common unit.

**RANGE/MICROWAVE/OVEN COMBINATION** – A kitchen appliance consisting of range/oven and microwave as one total unit built into a kitchen cabinet.

# KITCHENS

## KITCHEN COMPONENT DESCRIPTIONS (Cont'd)

**RANGE/OVEN** – A self-supporting stove with either a single or double oven.

**RANGE/OVEN (BUILT-IN)** – A range/oven unit that is built into a kitchen cabinet. May contain either a single or double oven.

**REFRIGERATOR** – A self-supporting refrigerator.

**REFRIGERATOR (UNDERCOUNTER)** – A refrigerator unit placed in or built into a kitchen cabinet, beneath the countertop.

**SINK (KITCHEN)** – A kitchen sink fixture with either one or two basins.

**TRASH COMPACTOR** – A trash compactor unit built into a kitchen cabinet.

**WALL CABINET** – A wall-mounted, wood or metal case consisting of shelves and doors.

# PLUMBING

Appliance Hookup
Drain
Faucets
Fixture Connections
Fixture Rough-in
Fountain - Decorative
Garbage Disposal
Hose Bib
Hot Water Heater
Piping
Pump - Circulating
Pump - Sump
Sink - Laundry
Sink - Wet Bar
Tubing - Copper
Water Filter
Water Softener
Well - Drill and Case
Well Pump
Well Water Tank

# PLUMBING

## TYPICAL PLUMBING LAYOUT AND PIPE SIZES

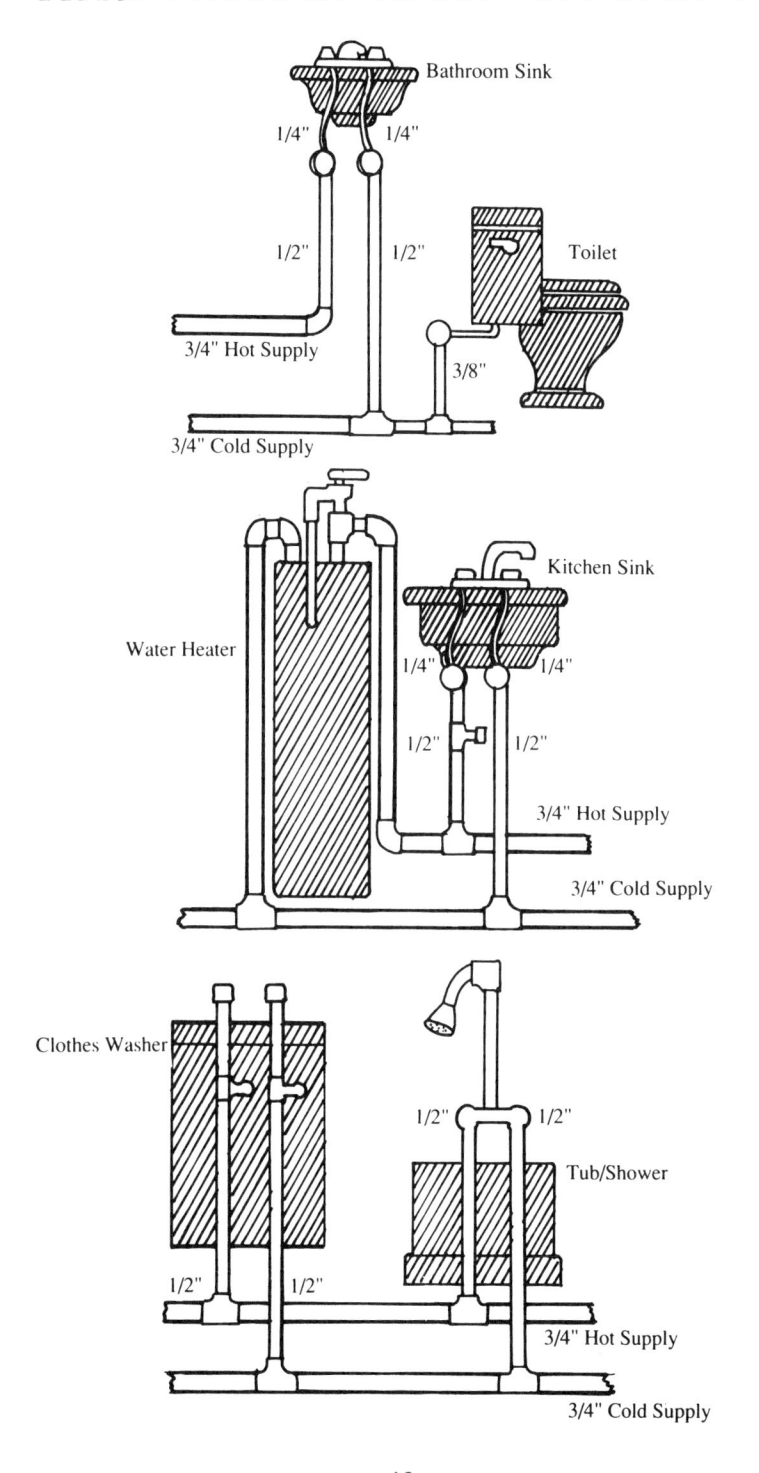

Bathroom Sink
1/4"  1/4"
1/2"  1/2"
Toilet
3/4" Hot Supply
3/8"
3/4" Cold Supply

Kitchen Sink
Water Heater
1/4"  1/4"
1/2"  1/2"
3/4" Hot Supply
3/4" Cold Supply

Clothes Washer
1/2"  1/2"
Tub/Shower
1/2"  1/2"
3/4" Hot Supply
3/4" Cold Supply

# PLUMBING

## PLUMBING COMPONENT DESCRIPTIONS

**APPLIANCE HOOKUP** – Piping connections necessary to provide gas or water to an appliance.

**CLEAN SEWER PIPE** – Cleaning of flow restrictions in a sewer pipe.

**DRAIN (FLOOR)** – An opening in a floor to drain water into a drain or sewer system.

**DRAIN (ROOF)** – A drain used on a roof to collect water.

**FAUCET (COMBINATION)** – A plumbing valve that combines hot and cold water through one outlet.

**FAUCET (DOUBLE)** – A plumbing valve with two water outlets.

**FAUCET (SINGLE)** – A plumbing valve with one water outlet.

**FIXTURE CONNECTION** – A device for joining together a plumbing fixture with the plumbing system.

**FIXTURE ROUGH-IN** – Installation of all parts of a plumbing connection that are completed prior to installation of a fixture.

**FOUNTAIN (DECORATIVE)** – A pump system whose fountain is of typical residential quality, made of concrete, plastic or fiberglass.

**GARBAGE DISPOSAL** – An electric device for grinding waste food prior to its entering the sewer pipe.

**HOSE BIB** – An exterior water faucet that is threaded to provide a connection for a hose.

**HOT-WATER HEATER** – Package equipment for heating water that uses gas or electricity.

**HOT-WATER HEATER (TANKLESS)** – A water heater whereby water is heated as it passes over heating elements.

**HOT-WATER TANK INSULATION** – Insulating material surrounding a hot-water tank.

**PIPING (CAST IRON)** – Piping made of cast iron.

# PLUMBING

## PLUMBING COMPONENT DESCRIPTIONS (Cont'd)

**PIPING (COPPER)** – Lightweight rigid copper piping joined by soldering.

**PIPING (GALVANIZED STEEL)** – Steel piping coated with zinc with fittings that are threaded.

**PIPING (PLASTIC)** – Piping made of a synthetic material (pvc).

**PUMP (CIRCULATING)** – A pump used to provide a continuous flow of water within a closed circuit.

**PUMP (SUMP)** – A pump used to drain the accumulated liquid from a receptacle.

**SERVICE SHUT-OFF VALVE** – A valve for a primary service line.

**SINK (LAUNDRY)** – A deep, wide sink usually of porcelain or steel which can have either one or two basins.

**SINK (WET BAR)** – A small basin usually used in a residential bar arrangement.

**TUBING (COPPER)** – Flexible copper piping joined by soldering, flaring, or compression fittings.

**WATER FILTER** – An in-line device used to trap sediment from the water supply.

**WATER SOFTENER** – An apparatus that chemically removes the calcium and magnesium minerals from a water supply.

**WELL (DRILL & CASE)** – A complete well. Includes engineering, setup, drilling, casing, sanitation and wellhead fixtures but excludes excavation and piping to carry water from the well to the house.

**WELL PUMP** – A submersible pump used to pump water from a well to a storage tank.

**WELL WATER PRESSURE TANK** – A water tank designed to pressurize a well water system.

# ROOFING

Asphalt - Hot Mopped
Built-up
Coping
Copper
Downspout
Elastomeric
Fascia
Felt Paper
Fiberglass
Flashing
Gravel Stop
Gutter
Insulation
Metal
Plastic Tile
Roll Roofing
Sheathing
Shingles
Skylight
Soffit
Tile - Clay
Tile - Concrete
Tile - Plastic
Tile - Slate
Wood Shakes

# ROOFING

## TYPICAL ROOF STYLES

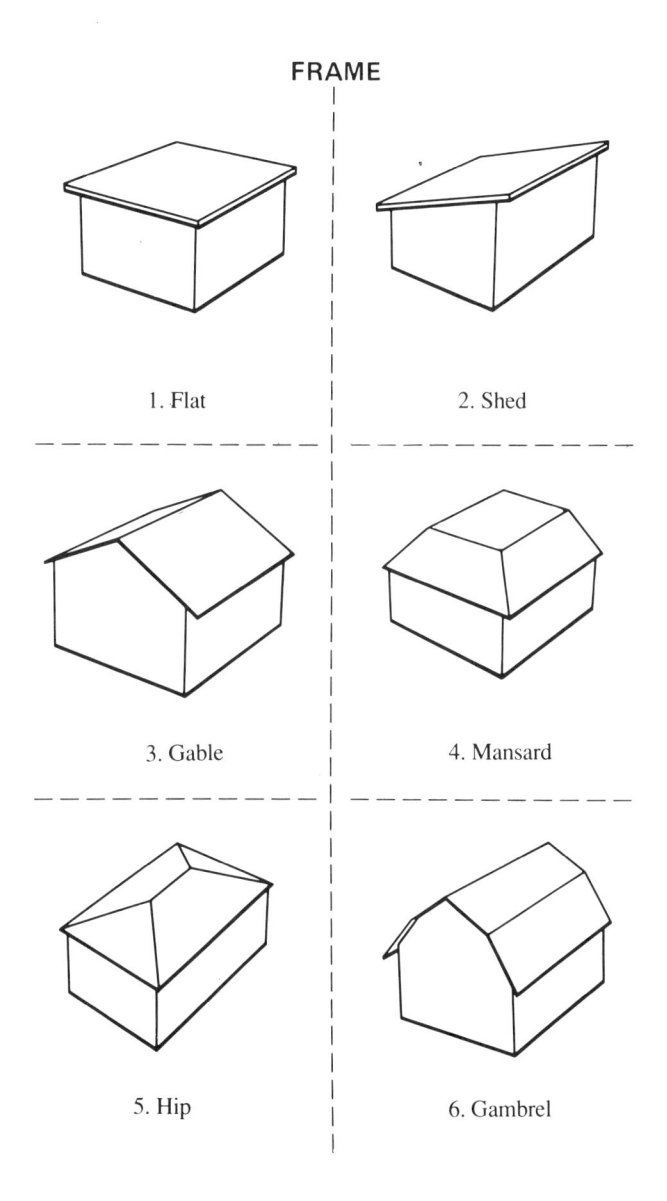

FRAME

1. Flat

2. Shed

3. Gable

4. Mansard

5. Hip

6. Gambrel

# ROOFING

## ROOF TERMINOLOGY

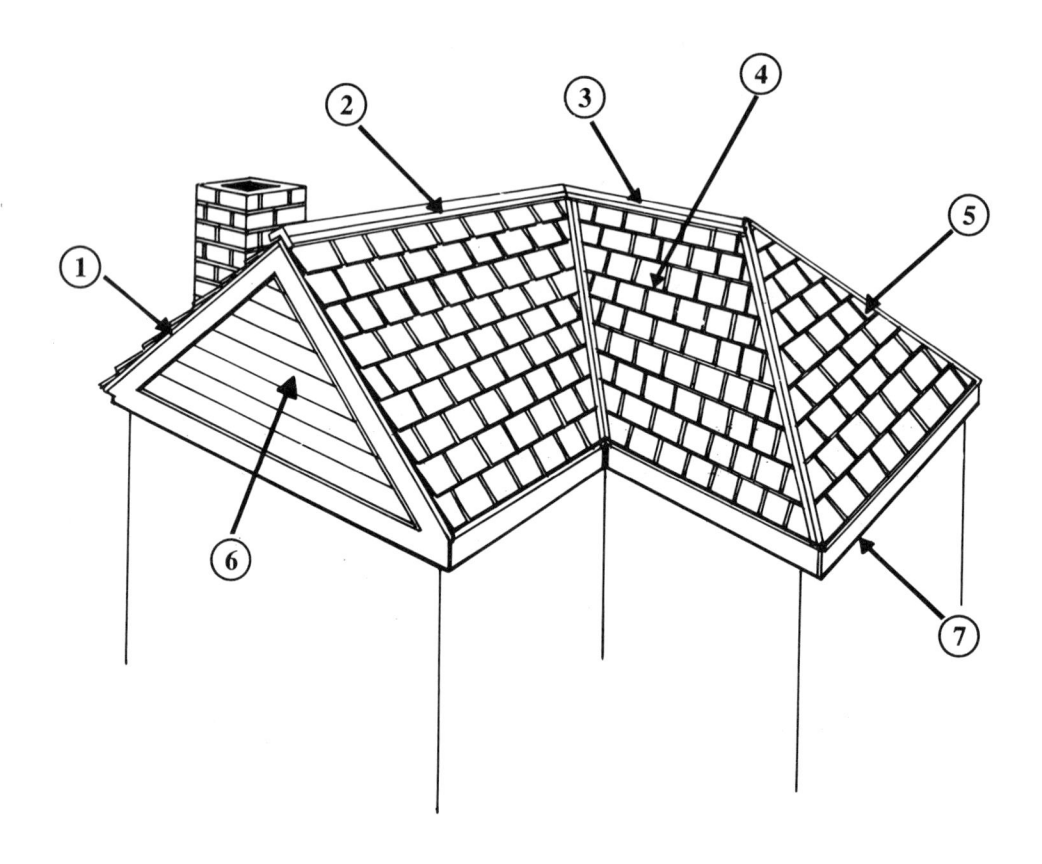

1. Rake
2. Ridge Vent
3. Ridge
4. Valley
5. Hip
6. Gable End
7. Eaves

# ROOFING

## TYPICAL ROOFING MATERIALS

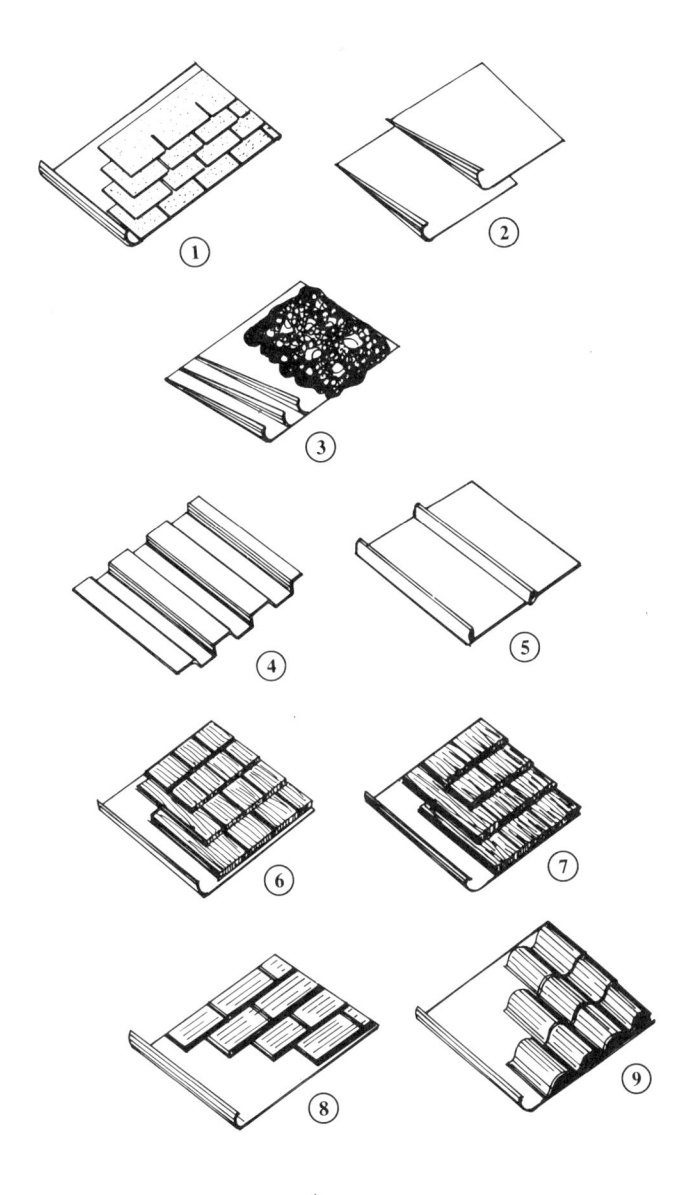

1. Asphalt Shingle (Low - $)
2. Roll (Low - $)
3. Built-up (Medium - $)
4. Preformed Metal (Low - $)
5. Standing Seam (Medium - $)
6. Wood Shingle (Medium - $)
7. Wood Shake (High - $)
8. Slate (High - $)
9. Spanish Tile (High - $)

55

# ROOFING

## ROOFING COMPONENT DESCRIPTIONS

**ASPHALT (HOT MOPPED)** – A protective roof covering comprised of a layer of hot asphalt.

**BUILT-UP** – A continuous roof covering of layers of saturated or coated felts that are alternated with layers of hot asphalt.

**CLAY TILE** – A roof covering of clay tiles, commonly called Spanish or mission tiles, approximately 13 1/4" x 9 3/4".

**CONCRETE TILE** – A roof covering of tile made from concrete.

**COPING** – A cap placed on top of a parapet wall to act as waterproofing.

**COPPER** – A roof covering of copper sheets with flat or standing seams.

**DOWNSPOUT** – A vertical pipe, often of sheet metal, used to channel water from a roof or gutter to the ground.

**ELASTOMERIC (SINGLE PLY)** – A roofing material having elastic properties that is capable of expanding or contracting with the surfaces to which the material is applied, without rupturing.

**FASCIA** – An exterior piece of trim that is placed at the end of roof rafters.

**FELT PAPER** – Asphalt-saturated or asphalt-coated felt applied over the roof deck or sheathing and under the roof cover.

**FIBERGLASS** – A protective roof covering comprised of a synthetic material.

**FIBERGLASS (CORRUGATED)** – A roof cover of corrugated fiberglass sheets.

**FLASHING** – A thin, continuous strip of metal, plastic, rubber or waterproofing membrane used to prevent the passage of water through a joint.

**GRAVEL STOP** – A metal piece placed along the edge of a gravel-covered roof to prevent the gravel from washing over the roof edge.

**GUTTER** – A channel of metal, plastic, or wood to catch and carry off rain water from a roof to the downspout.

# ROOFING

## ROOFING COMPONENT DESCRIPTIONS (Cont'd)

**INSULATION (BATT)** – A flexible blanket-type thermal insulation commonly used as insulation between rafters or joists in frame construction.

**INSULATION (RIGID)** – A rigid insulation; usually boards of polystyrene, mineral fiberglass, glass fiber, etc., applied on or below the roof sheathing.

**METAL** – A roof covering of either corrugated or shallow rib steel or aluminum.

**PLASTIC TILE** – A roof covering of tile made from plastic.

**ROLL ROOFING (COMPOSITION)** – A mineral-surfaced roof covering manufactured in rolls.

**SHEATHING** – A covering of plywood sheets on roof rafters.

**SHINGLES (COMPOSITION)** – A roof covering of composition shingles of various weights. No felt underlayment is included.

**SKYLIGHT** – A glazed opening in a roof to allow in natural light.

**SLATE TILE** – A roof covering of cut slate tiles.

**SOFFIT** – An exposed undersurface of a roof overhang.

**WOOD SHAKES** – A roof covering of thick wood shakes with an average exposure.

**WOOD SHINGLES** – A roof covering of standard size wood shingles with an average exposure.

# SITEWORK

Barbecue
Curbs
Fencing
Flagpole
Landscaping
Lawn Sprinkler
Lighting
Mailbox
Paving - Asphalt
Paving - Brick
Paving - Concrete
Paving - Flagstone/Tile
Paving - Wood
Ramps
Septic Tank
Steps
Swimming Pool/Spa
Wall - Brick
Wall - Retaining
Wall - Stone
Wall - Block/Concrete
Wall - Block/Ornamental
Wood Decks

# SITEWORK

## TYPICAL WINDBREAK LAYOUT

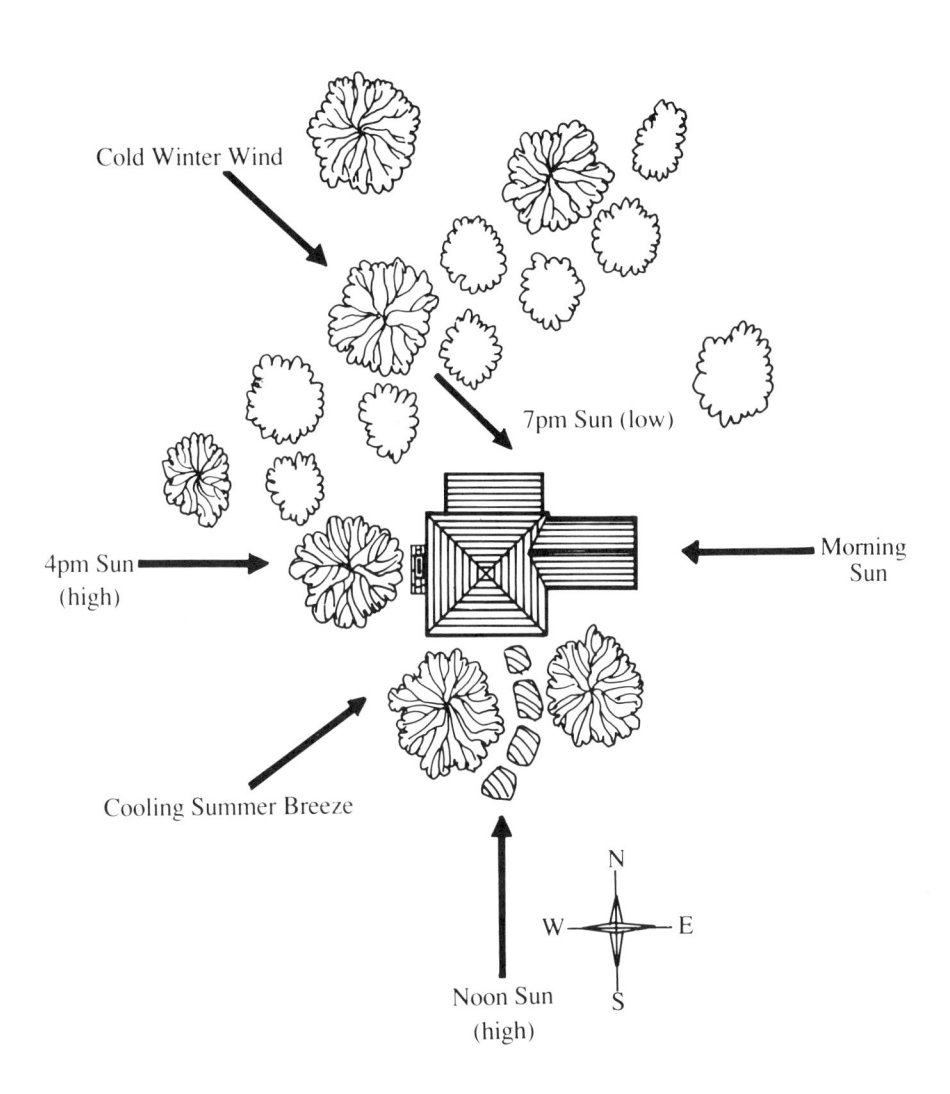

Cold Winter Wind

7pm Sun (low)

4pm Sun (high)

Morning Sun

Cooling Summer Breeze

Noon Sun (high)

N / W / E / S

59

# SITEWORK

## SITEWORK COMPONENT DESCRIPTIONS

**ASPHALT PAVING** – A paving consisting of asphalt cement.

**BARBECUE (BUILT-IN)** – A masonry-constructed box and chimney used for outdoor cooking.

**BLOCK WALLS** – A wall constructed of concrete block.

**BRICK PAVING** – A vitrified brick, laid flat or on edge, especially suitable for use in pavements.

**BRICK WALLS** – A wall constructed of brick.

**CHAIN-LINK FENCE** – Fencing consisting of tubular end, corner and pull posts, top and bottom rails or tension wires, and faced with a mesh fabric.

**CHAIN-LINK GATE** – A chain-link gate including posts, rails and tension wires.

**CONCRETE PAVING** – A paving consisting of concrete.

**CURBS (CONCRETE)** – A raised rim of concrete forming the edge of a street.

**FLAGPOLE** – A pole, base or setting, pulleys and rope that make up a flagpole.

**FLAGSTONE/TILE PAVING** – A surface of flat stone or tile units laid on a concrete base.

**HOT TUB** – A therapeutic bath with temperatures that range up to 104°. It may include a whirlpool feature.

**LANDSCAPING** – Lawn, flowers, immature trees and bushes and other flora constituting a landscape.

**LAWN SPRINKLER CONTROL** – A lawn sprinkler control device with an automatic clock timer.

**LAWN SPRINKLER HEAD** – That portion of piping connecting a water supply to a sprinkler system.

# SITEWORK

## SITEWORK COMPONENT DESCRIPTIONS (Cont'd)

**LIGHTING** – A yard-lighting fixture installed either on a post or on the ground.

**MAILBOX (POST TYPE)** – A standard mailbox mounted on a post.

**ORNAMENTAL IRON GATE** – A gate constructed of decorative iron.

**RAMPS (CONCRETE)** – A slope constructed of reinforced concrete, joining two levels.

**REDWOOD FENCE** – A fence consisting of pieces of soft redwood.

**REFINISH SWIMMING POOL** – Removal of a damaged swimming pool surface finish, and the application of a new finish.

**REPAIR SWIMMING POOL CRACKS** – Filling in and surfacing over the cracks.

**REPAIR SWIMMING POOL TILES** – Replacement of damaged tiles with new tiles.

**REPAIR WOOD FENCE** – The repair operation of re-aligning and bracing a wood fence that has fallen or been pushed down.

**REPLASTER SWIMMING POOL** – Replastering of a damaged swimming pool surface.

**RETAINING WALL** – A freestanding or laterally braced wall that bears against an earth or other fill surface.

**SEPTIC DISTRIBUTION BOX** – A box in which the effluent from a septic tank ensures equal distribution to each individual line of the disposal field.

**SEPTIC LEACHING LINES** – Lines used to distribute sewage effluent throughout a leach field.

**SEPTIC TANK** – A watertight, covered receptacle designed and constructed to receive the discharge of sewage, separate solids from liquids, digest organic matter, store digested solids through a period of detention and allow the clarified liquids to discharge for final disposal.

# SITEWORK

## SITEWORK COMPONENT DESCRIPTIONS (Cont'd)

**SPA** – A spa adjoined to or separate from a swimming pool. Generally includes benches and aerators.

**SPLIT-RAIL FENCE** – A fence made by connecting horizontal rails of split wood to posts.

**STEPS** – Steps built with concrete or clay bricks.

**STONE WALL** – A wall made of stone found in the local area.

**SWIMMING POOL** – A swimming pool constructed of either concrete, sprayed gunite, fiberglass or plastic-lined frame.

**SWIMMING POOL COVER** – A cover for the swimming pool that is placed over the pool by hand.

**SWIMMING POOL DIVING BOARD** – Board, springs and other associated hardware related to a diving board.

**SWIMMING POOL FILTER** – An apparatus for clarifying pool water.

**SWIMMING POOL HEATER** – A device designed to heat pools.

**SWIMMING POOL LADDER** – A stainless steel pool ladder.

**WOOD BOARD FENCE** – A fence of continuous wood boards that are butted together.

**WOOD DECKS** – A flat, open platform constructed of wood. Includes piers and posts, but excludes railings.

**WOOD PAVING** – A durable flat surface of wood, 2" x 4", laid flat or on edge and set on a foundation of subgrade soil.

**WOOD PICKET FENCE** – A fence formed of a series of wood pickets.

**WROUGHT IRON FENCE** – A fence made of decorative iron.

# SPECIALTIES/ EQUIPMENT

Blinds
Broom Closet
Closet Pole
Clothes Dryer
Clothes Washer
Draperies
Drapery Rod
Drapery Track
Fireplaces
Greenhouse
Hot Tub/Spa
Mailbox - Wall Type
Patio Enclosure
Railing
Sauna Room
Shades
Shelving
Shutters - Interior
Spiral Stairs
Wardrobe
Workbench

# SPECIALTIES/EQUIPMENT

## SPECIALTIES & EQUIPMENT COMPONENT DESCRIPTIONS

**BLINDS** – Blinds made of thin horizontal or vertical slats.

**BROOM CLOSET** – A metal or wood case, for broom storage.

**CLOSET POLE** – A long rod of wood or metal, used to hang garments in a closet.

**CLOTHES DRYER** – A self-supporting clothes dryer.

**CLOTHES WASHER** – A self-supporting clothes washer.

**DESK (BUILT-IN)** – A wood or metal desk, permanently affixed to a structure.

**DRAPERIES** – Material used to cover or decorate windows.

**DRAPERY ROD** – An ornamental horizontal pole, on which to hang drapery.

**DRAPERY TRACK** – A horizontal device used to support, and allow the opening and closing of, window drapes.

**FIREPLACE CHIMNEY (MASONRY)** – A vertical masonry structure, containing one or more flues.

**FIREPLACE HEARTH (RAISED)** – A hearth that is raised above floor level.

**FIREPLACE LOG LIGHTER** – A device for igniting wood logs in a fireplace.

**FIREPLACE (MASONRY)** – A masonry box at the base of a chimney, usually an open recess in a wall.

**FIREPLACE (MASONRY W/HEATILATOR)** – A masonry fireplace with the extra cost of a heatilator system.

**FIREPLACE (PREFAB)** A box at the base of a chimney, usually an open recess in a wall, constructed of metal.

**GREENHOUSE** – A glass-enclosed, prefabricated greenhouse.

**HOT TUB/SPA** – A therapeutic bath with temperatures that range up to 104°. Either may include a whirlpool feature.

# SPECIALTIES/EQUIPMENT

## SPECIALTIES & EQUIPMENT COMPONENT DESCRIPTIONS
## (Cont'd)

**MAILBOX (WALL TYPE)** – A box recessed in, or attached to, the outer wall of a residence.

**PATIO ENCLOSURE** – An outside area enclosure, attached to a house, made of mesh/screened or glass/acrylic walls.

**RAILINGS** – Wall-mounted or freestanding, metal or wood railings.

**SAUNA HEATING UNIT** – A unit that supplies heat for a sauna.

**SAUNA ROOM** – A hot-air bath.

**SHADES** – A roller-type window shade.

**SHELVING** – Metal or wood shelves, used for storage.

**SHUTTERS (INTERIOR)** – Interior window covers, with movable slats.

**SPIRAL STAIRS** – A flight of aluminum, steel, or wood steps, circular in design, whose treads wind around a central newel.

**STOVE (WOOD/COAL BURNING)** – A self-contained wood or coal-burning stove.

**WARDROBE** – A built-in wood or metal storage unit, for clothing.

**WORKBENCH** – Metal or wood worktable.

# WALL FINISHES

Block - Concrete
Brick
Gypsum Board
Hardboard
Insulation
Mirror
Molding
Paint
Paneling - Plywood
Paneling - Solid
Plaster
Stain
Studs
Tile - Ceramic
Tile - Metal
Tile - Mirror
Tile - Plastic
Trim
Veneer - Brick
Wainscot
Wallpaper

# WALL FINISHES

## DRYWALL FINISH SYSTEM

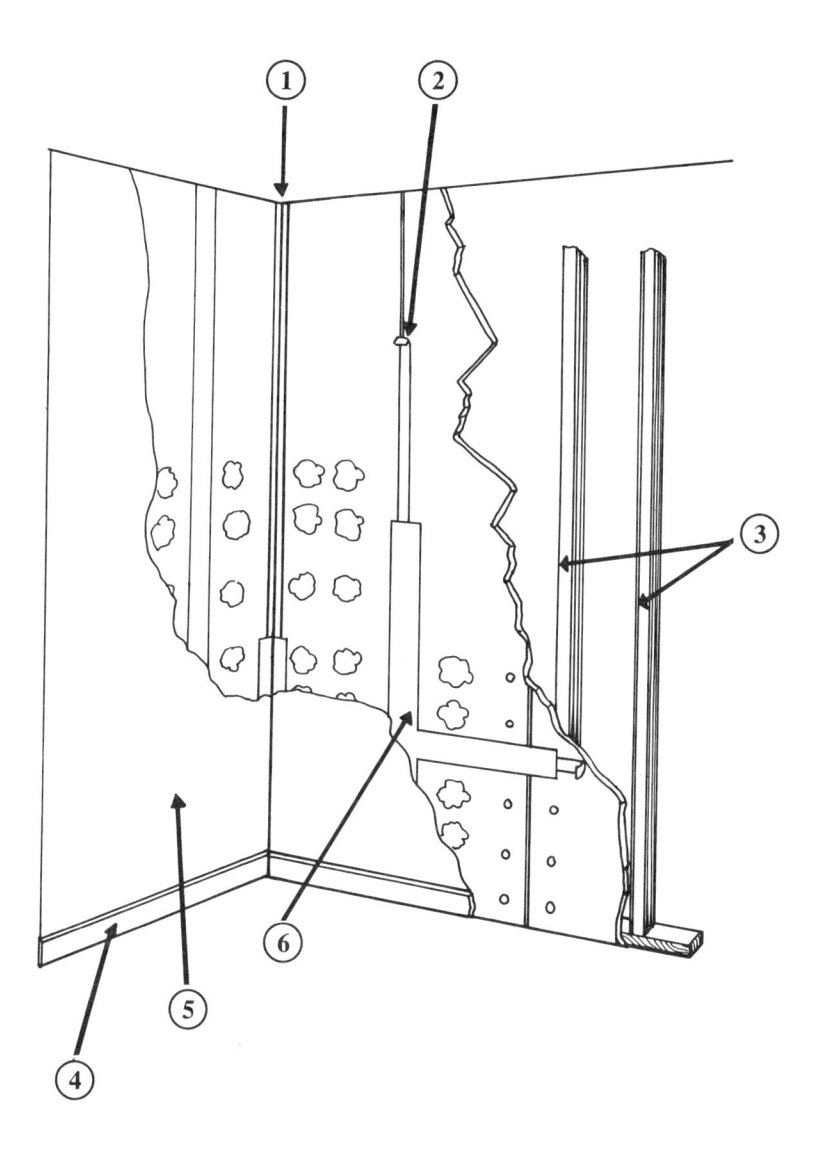

1. Corners
2. Tape Seams
3. Studs
4. Trim
5. Paint or Wallpaper
6. Finish

# WALL FINISHES

## WALL FINISH COMPONENT DESCRIPTIONS

**BLOCK (CONCRETE)** – An interior concrete, or masonry block, non-load bearing wall. Normal thickness 4" to 6".

**BRICK (SOLID)** – An interior brick wall. May be load or non-load bearing.

**BRICK VENEER** – An interior wall covering of face brick, laid against a frame or masonry wall.

**BRICK VENEER (SIMULATED)** – An interior wall covering, that has the appearance of brick veneer.

**CAULKING** – Installation of a resilient mastic compound used to seal cracks, fill joints, prevent leakage and/or provide waterproofing.

**CERAMIC TILE** – An interior wall covering of ceramic tile units.

**GYPSUM BOARD** – 1/2" to 5/8" wallboards, also known as sheetrock or drywall.

**GYPSUM BOARD (WATER RESISTANT)** – A wallboard treated to resist the passage of water and moisture.

**HARDBOARD** – An interior wall covering of flush hardboard panels.

**INSULATING BOARD** – An interior wall covering of prefinished insulating boards.

**INSULATION (BATT)** – A flexible blanket or roll-type thermal insulation placed between studs in frame construction.

**INSULATION (RIGID)** – A structural building board, applied to walls to resist heat transmission.

**LATH ONLY** – A thin metal mesh, or strips of narrow wood, fastened to framing in order to provide a base for plaster.

**METAL TILE** – An interior wall covering of metal tile units.

**MIRROR** – An interior wall covering of reflective glass wall panels.

**MIRROR TILE** – An interior wall covering of reflective glass tile units.

# WALL FINISHES

## WALL FINISH COMPONENT DESCRIPTIONS (Cont'd)

**MOLDING (BASE)** – A horizontal strip, typically affixed to the bottom of the wall, to cover the flooring-to-wall material transition joint.

**MOLDING (CEILING)** – A horizontal strip, typically affixed to the top of the wall, to cover the ceiling-to-wall material transition joint.

**MOLDING (CHAIR RAIL)** – A horizontal strip, usually of wood, affixed to the wall at a height that prevents backs of chairs from damaging the wall surface.

**MOLDING (CORNER)** – A vertical strip, used to protect and trim an external angle, or two intersecting surfaces.

**MOLDING (PANEL EDGE)** – A metal or wood vertical piece, installed over the butt joint of two panels.

**PAINT** – A colored, adhesive coating which can be applied over a variety of surfaces for decoration and protection.

**PANELING (PLYWOOD)** – An interior wall covering of plywood paneling, with a softwood or hardwood surface.

**PANELING (SOLID)** – An interior wall covering of hardwood or softwood panels.

**PLASTER (ONLY)** – Three coats of gypsum plaster, applied to a wall surface.

**PLASTER & LATH** – An interior wall covering of plaster, with a metal lath backing.

**PLASTER (THINCOAT)** – An interior wall finish covering, comprised of a thin application of plaster.

**PLASTIC TILE** – An interior wall covering of plastic tile units.

**REGROUT TILE** – Reapplying cement mortar between tile joints on a wall surface.

**REPAIR PLASTER** – The repair of a plaster wall, such as repairing minor cracks, peeling, etc.

# WALL FINISHES

## WALL FINISH COMPONENT DESCRIPTIONS (Cont'd)

**REPOINT MASONRY** – The removal of mortar from between the joints of masonry units, and replacing it with new mortar.

**STAIN** – A wax base stain, applied as a protective coating or polish on interior wood.

**STONE VENEER (IMITATION)** – An interior wall covering that has the appearance of stone.

**STUDS** – An interior wall, constructed using wood studs.

**TRIM** – Visible woodwork or moldings of a room, such as baseboards, cornices, casings, etc.

**WAINSCOT** – A decorative or protective facing, applied to the lower portion of an interior partition or wall, such as panels, usually reaching a height of three to four feet.

**WALLPAPER** – A wall covering of quality paper or vinyl.

**WALLPAPER (GRASS CLOTH)** – A wall covering of woven grass cloth.

**WALLPAPER (MURAL)** – A wall covering of paper or vinyl on which a painting has been rendered.

# WINDOWS

Awnings
Glass
Hardware
Screens
Security Grille
Security Mesh
Shutters
Trim - Exterior
Trim - Interior
Windows

# WINDOWS

## WINDOW UNIT TYPES

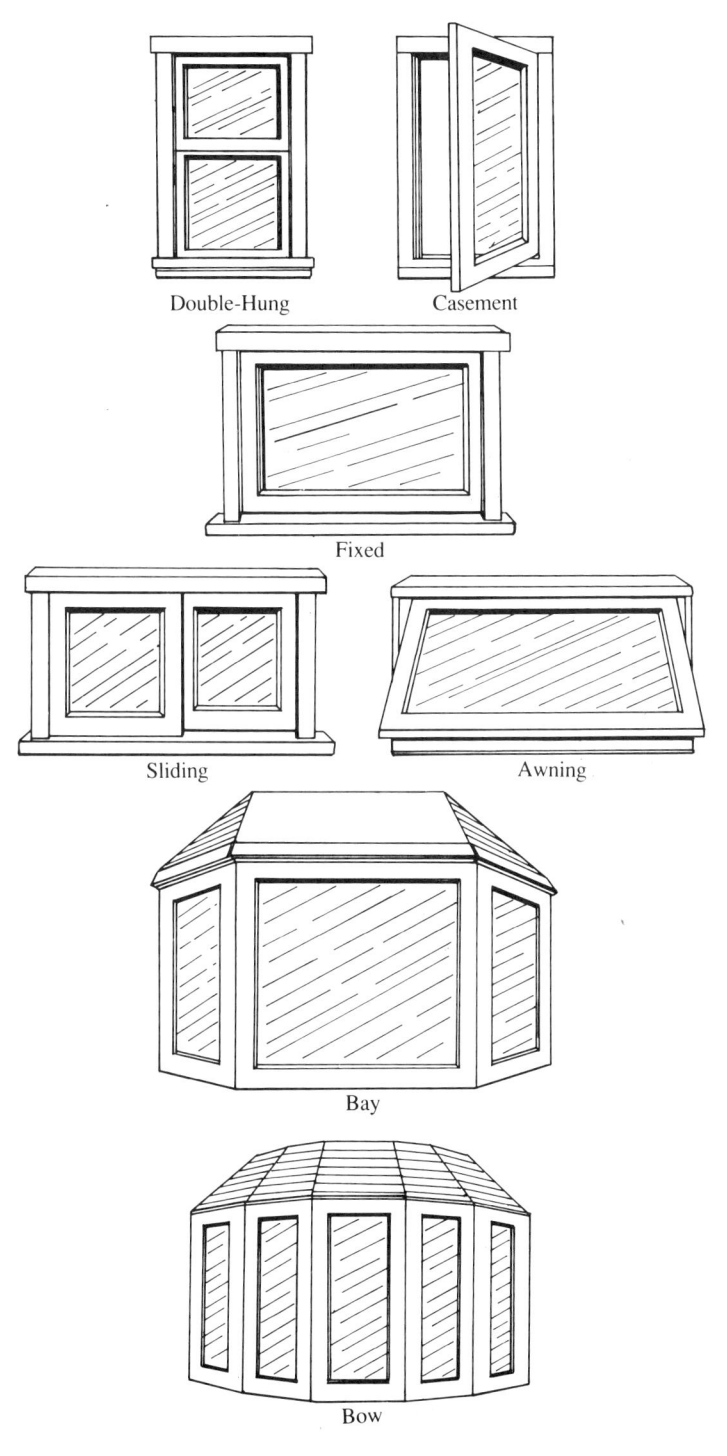

Double-Hung    Casement

Fixed

Sliding    Awning

Bay

Bow

69

# WINDOWS

## WINDOW TERMINOLOGY

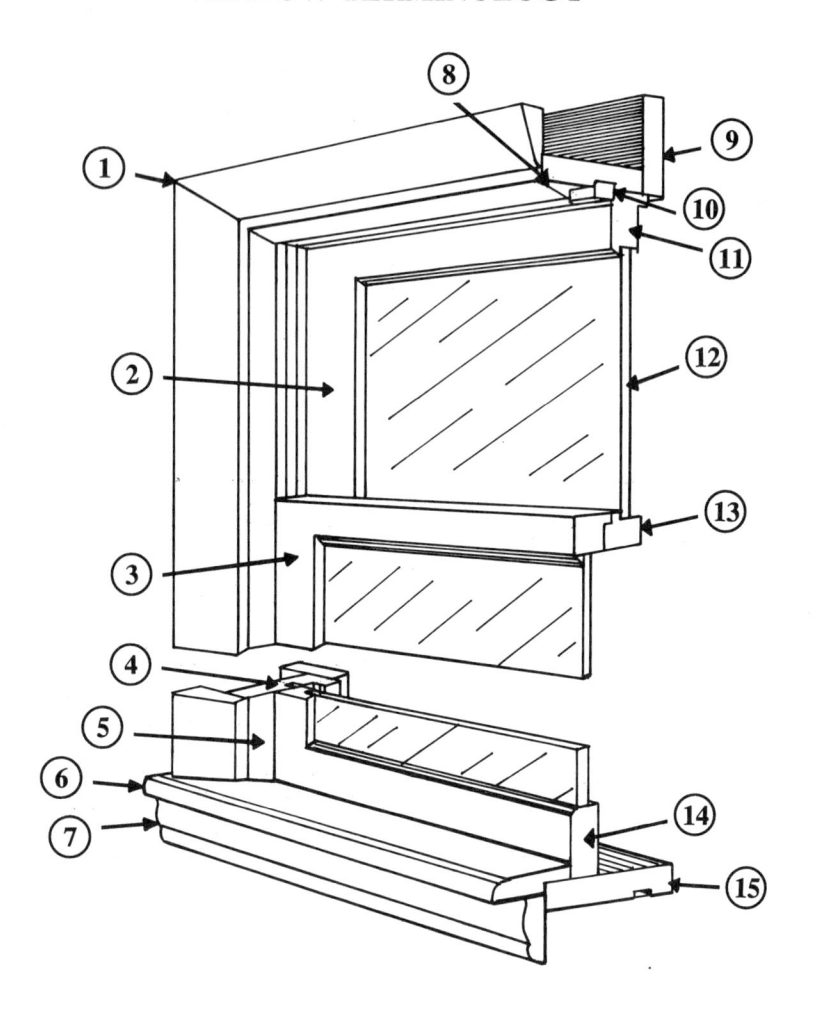

1. Inside Casing
2. Upper Sash
3. Lower Sash
4. Jamb
5. Stop
6. Stool
7. Apron
8. Stop
9. Outside Casing
10. Parting Stop
11. Rail
12. Glazing
13. Check Rail or Meeting Rail
14. Rail
15. Sill

# WINDOWS

## WINDOW COMPONENT DESCRIPTIONS

**ACRYLIC/PLEXIGLAS** – The material and installation of acrylic/plexiglas in a single frame.

**AWNINGS** – A rooflike covering of canvas or aluminum over a window.

**AWNING WINDOWS** – A sashed window that can be tilted outward.

**CASEMENT WINDOWS** – A sashed window that swings open from the side.

**COMBINATION WINDOWS** – A framed window that is equipped with removable or interchangeable screen and storm glass sections.

**DOUBLE-HUNG WINDOWS** – A sashed window having two vertically sliding sashes. Window trim is not included.

**FIXED WINDOWS** – A window which does not open.

**GREENHOUSE WINDOWS** – A metal-framed window extending approximately one to two feet outside a window opening.

**HARDWARE** – Items included in window hardware range from a variety of window locks to window handles.

**INSULATING GLASS** – The material and installation of insulation glass in a single frame.

**JALOUSIE** – A window consisting of glass louvers which pivot simultaneously in a common frame.

**JALOUSIE GLASS (ONLY)** – The material and installation of a series of overlapping, horizontal glass louvers which pivot simultaneously in a common frame.

**OPAQUE GLASS** – The material and installation of a glass that transmits light but cannot be seen through.

**PAINT** – A colored, adhesive coating which can be applied over a variety of surfaces for decoration and protection.

**PLATE GLASS** – The material and installation of 1/4" plate glass in a single frame.

# WINDOWS

## WINDOW COMPONENT DESCRIPTIONS (Cont'd)

**SCREENS** – A window screen whose frame is of wood or metal.

**SECURITY GRILLES** – A grating or openwork barrier of strong metal to protect an opening.

**SECURITY MESH** – Heavy mesh screening placed on the exterior of a window for added security.

**SHUTTERS (LOUVER)** – Exterior shutters with movable louvers.

**SHUTTERS (PLAIN)** – Exterior plain shutters, with or without fixed louvers.

**SINGLE GLASS** – The material and installation of one 1/8" window pane in a single frame.

**SINGLE-HUNG WINDOWS** – A framed window that is hung and balanced so that only the lower sash opens.

**SKYLIGHT GLAZING** – A glazed opening in a roof.

**SLIDING WINDOWS** – A sashed window that opens or closes horizontally along a fixed track.

**STORM WINDOWS** – An auxiliary window, usually placed on the outside of an existing window.

**TEMPERED GLASS** – The material and installation of tempered glass in a single frame.

**TRIM** – Any visible member, usually of wood, around the exterior perimeter of the window frame.

**WIRED GLASS** – The material and installation of wired glass in a single frame.

# USEFUL INFORMATION

House Style Descriptions
Weights & Measures
Climate Classification Key
Insulation Requirements

# USEFUL INFORMATION

## TYPICAL HOUSE STYLES

### ONE STORY:
One Story residences have one level of living area. The roof structure has a medium slope. The attic space is limited and is not intended for living area.

### ONE AND ONE HALF STORY:
One and One Half Story residences have two levels of living area. Characterized by a steep roof slope and dormers, the area of the upper level, whether finished or unfinished, is usually 40% to 60% of the lower level.

### TWO STORY:
Two Story residences have two levels of finished living area. The area of each floor is approximately the same. The roof structure has a medium slope. The attic space is limited and is not designed for usable living area.

### TWO STORY BI-LEVEL:
Two Story Bi-level residences have two levels of living area, but unlike a conventional Two Story, the lower level, which may be partially below grade, is partially unfinished. A distinguishing characteristic is its split-foyer entry.

### TWO AND ONE HALF STORY:
Two and One Half Story residences have three levels of living area. Also has a steep roof slope with dormers; the area of the third floor, whether finished or unfinished, is usually 40% to 60% of the second floor.

# USEFUL INFORMATION

## TYPICAL HOUSE STYLES (Cont'd)

**SPLIT LEVEL:**
Split level residences have three levels of finished living area: lower level, intermediate level and upper level. The lower level is immediately below the upper level as in a two story. The intermediate level, adjacent to the other levels, is built on a grade approximately four feet higher than that of the lower level.

**MOBILE / MANUFACTURED HOUSING:**
Often referred to as mobile homes, these structures, whether on a permanent or semipermanent foundation, have a steel undercarriage as a necessary structural component.

**MULTIPLES**:
Multiples, often referred to as apartments, are multifamily residences, intended for permanent habitation, and are three stories or less.

**TOWN HOUSES AND DUPLEXES:**
Both Town Houses and Duplexes are single family, attached residences. They do not have other units above or below, do not have more than two walls that are common with adjacent units and always have individual exterior entries.

# USEFUL INFORMATION

## TYPICAL HOUSE STYLES (Cont'd)

### URBAN ROW HOUSES:

Urban Row Houses are single family residences and can be either attached or detached. Unlike Town Houses, Urban Row Houses are usually individually built, with adjacent units not sharing common structural systems (i.e., roof, foundation, etc.). A distinguishing characteristic is that the living area is entirely on the second level. The ground floor, sometimes referred to as the basement level, usually contains the garage and utility area.

## TYPICAL GARAGE STYLES

### DETACHED GARAGES:

Detached Garages are freestanding buildings with independent structural systems (i.e., foundation, roof, etc.).

### ATTACHED GARAGES:

Attached Garages share a common wall with the residence.

### BUILT-IN GARAGES:

Built-In Garages have living area both adjacent to and above.

### BASEMENT GARAGES:

Basement Garages have living area above and have two to three walls below grade.

# USEFUL INFORMATION

## CARPORTS

A Shed or Flat Roof has a two-dimensional roof structure.

A Gable Roof has a three-dimensional, trussed roof system and is usually an extension of the residence roof structure.

# USEFUL INFORMATION

## HEIGHT MEASUREMENT TECHNIQUE

The total height of a building may be measured by dropping a tape from the roof of the building, by measuring down through a stairwell or similar vertical opening, or by measuring ceiling heights and allowing for the thickness of the floors. Sometimes the height is more easily estimated by standing back from the building and sighting using the following formula:

$H = A \times \dfrac{h}{a}$

$A$ = Distance from eye to wall

$a$ = Distance from eye to ruler

$h$ = Distance on ruler

$H$ = Height of wall

For example, substituting in the aforementioned formula:

$H = 90' \times \dfrac{21}{27}$

$H = 70'$

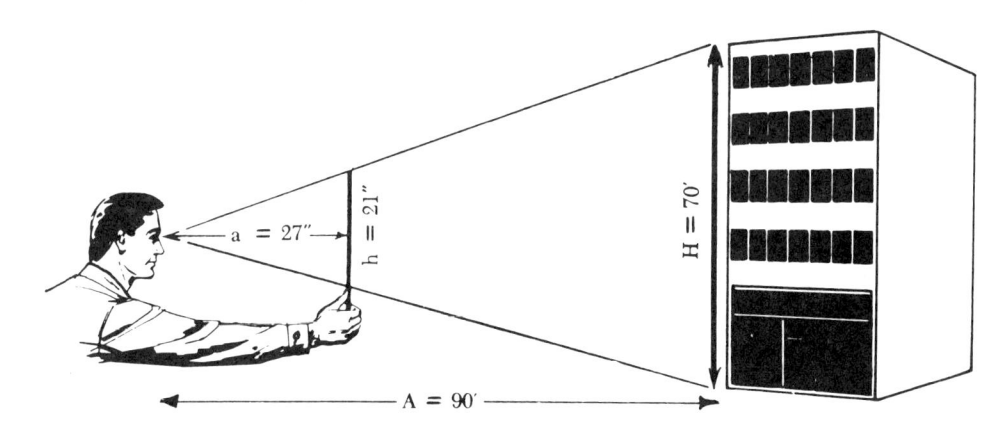

# USEFUL INFORMATION

## DETERMINING AVERAGE STORY HEIGHT

When figuring the average story height of a building with a balcony or mezzanine, these added interior items should be disregarded.

In the case of high-pitched roofs, some adjustment should be made for the gable ends and the large roof area. Usually, it is sufficiently accurate to add one-half the vertical distance from the eave line to the ridge to obtain an effective wall height for adjustment. This can also be used with an A-frame, where the effective wall height would be one-half the distance from the floor to the ridge.

Normally, the average story height of the building is the distance from the ground or top of the basement wall to the eaves, divided by the number of stories.

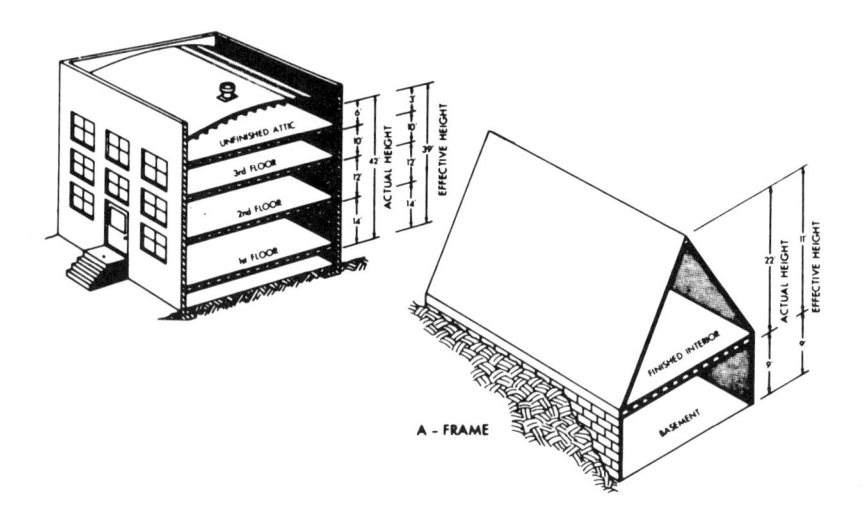

## MEASURING FLOOR AREAS

Use a steel or metallic tape with a hook on one end for your measuring. Metallic tapes are cloth tapes with wire in them to keep their length uniform. Plain cloth tapes are unreliable as they tend to shrink or stretch. Caution should be used with both metallic and steel tape around electricity. If a wheel device is used, it should be closely checked and calibrated.

Draw each line to scale and go completely around the building, bringing your lines back to the point of beginning. Errors in scale or measurement will then show up while you are in the field.

# USEFUL INFORMATION

## COMPUTING AREAS

Floor areas are based on the outside dimensions of the building. Most buildings can be broken down into rectangles and the area of each rectangle computed with no difficulty. Other shapes can be computed as shown below.

### Parallelogram

A quadrilateral having its opposite side parallel.

AREA = b x h

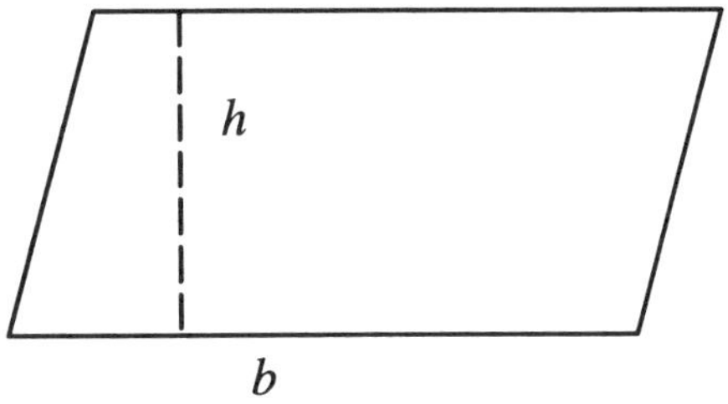

The altitude (h) of a parallelogram or trapezoid is the perpendicular distance between the parallel sides.

### Trapezoid

A quadrilateral having two and only two sides parallel.

AREA = h x 1/2 (b + c)

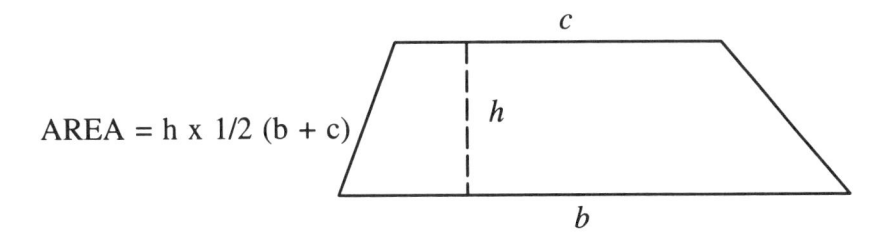

# USEFUL INFORMATION

## COMPUTING AREAS (Cont'd)

### Triangle

A three-sided polygon.

AREA = 1/2 b x h

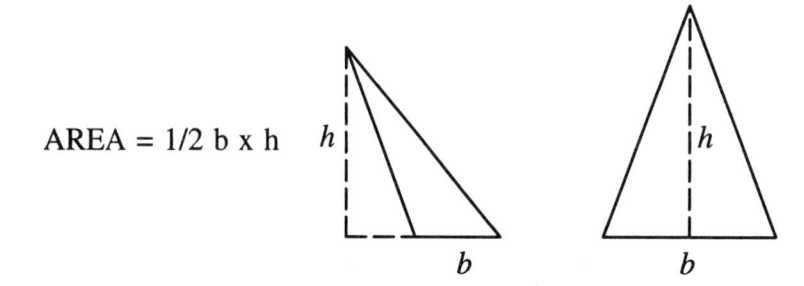

The altitude (h) of a triangle is the perpendicular distance from any vertex to the opposite side or its extension.

### Irregular Polygons

The area of irregular polygons can be determined by dividing the area into the shapes on the previous page and adding the area of the parts.

AREA = 1/2 a x b

+ c x 1/2 (b + d)

+ 1/2 e x d

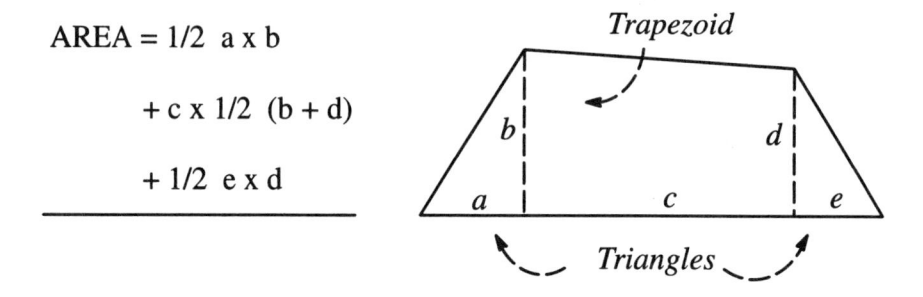

*Trapezoid*

# USEFUL INFORMATION

## COMPUTING AREAS (Cont'd)

### Trapezium

A quadrilateral having no two sides parallel. The area of a trapezium can only be determined by dividing the figure into triangles, parallelograms and/or trapezoids and totalling the individual areas.

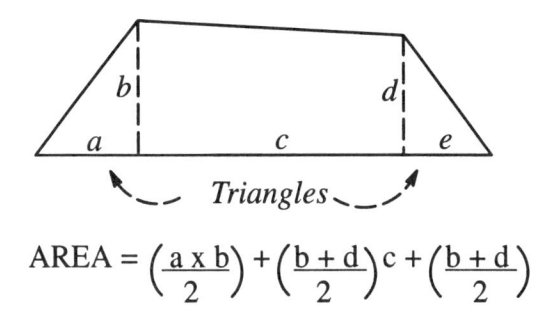

$$AREA = \left(\frac{a \times b}{2}\right) + \left(\frac{b+d}{2}\right)c + \left(\frac{b+d}{2}\right)$$

### Properties of a Circle

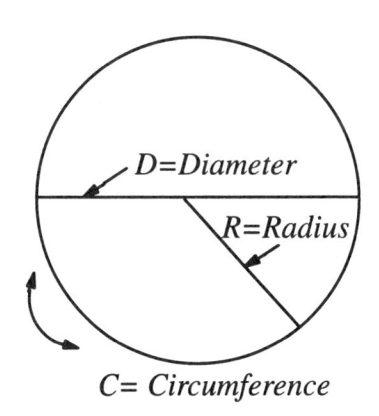

Area   = $D^2$ x .7854
    = $R^2$ x 3.1416
    = $C^2$ x .07958
    = R x 2

Diameter = C x .3183

Circumference  = D x 3.1416
        = R x 6.283185

Radius      = D ÷ 2
        = C x .159155

D = Diameter
R = Radius
C = Circumference

# USEFUL INFORMATION

## CLIMATE CLASSIFICATION KEY

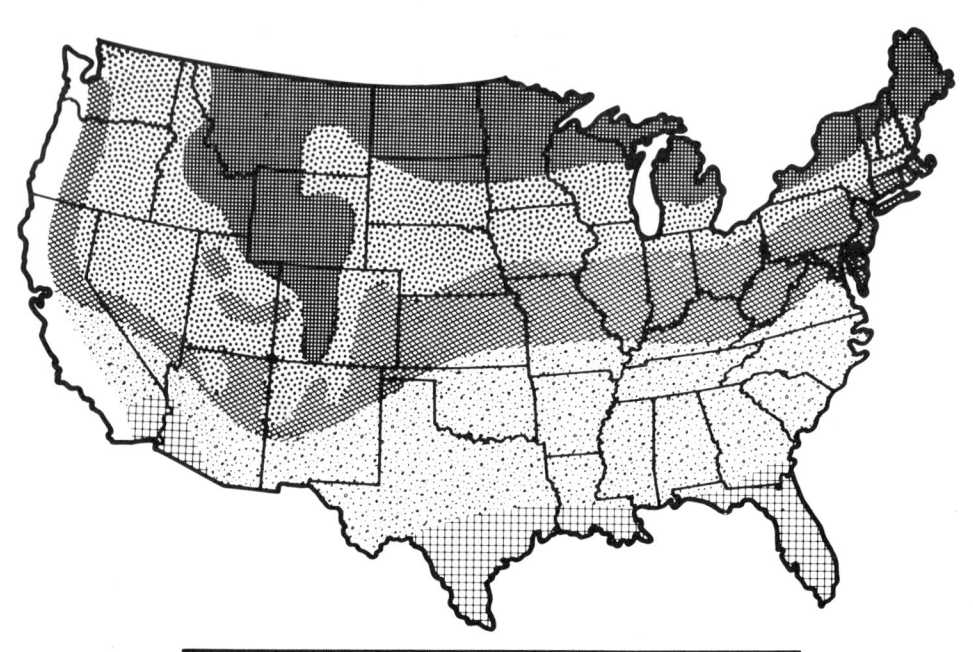

| CLIMATE CLASSIFICATION: | MILD | | MODERATE | | EXTREME | |
|---|---|---|---|---|---|---|
| TYPICAL INSULATION (R-VALUE) CEILING: | R-19 | R-26 | R-26 | R-30 | R-33 | R-38 |
| WALL: | R-11 | R-13 | R-19 | R-19 | R-19 | R-19 |
| FLOOR: | R-11 | R-11 | R-13 | R-19 | R-22 | R-22 |

# USEFUL INFORMATION

## INSULATION REQUIREMENTS

The following table lists the typical thickness required: 1/2" at a designated value for fiberglass or mineral wool insulation which is used in residential construction for the ceiling, wall and floor areas. Rockwool is typically 1/2" thinner than fiberglass at the same R-value. R-values are averages of unfaced, foil-faced and kraft paper-faced insulation when available.

### Ceilings

Fiberglass batt or blanket insulation

| | |
|---|---|
| R-13 | One 3 5/8" batt |
| R-19 | One 6 1/2" batt |
| R-26 | Two 3 5/8" batts |
| R-30 | One 6 1/2" batt & one 3 1/2" batt |
| R-35 | One 7" Batt & 3 5/8" batts |
| R-38 | Two 6 1/2" batts |

Loose fill wood & fiberglass batts or blankets:

| | |
|---|---|
| R-19 | One 7 1/2" wool fill or 6 1/2" batts |
| R-26 | 2 1/2" wool fill & 6 1/2" batts |
| R-30 | 4 1/2" wool fill & 6 1/2" batt |
| R-38 | 7 1/2" wool fill & 6 1/2" batt |

### Walls

| | |
|---|---|
| R-5.5 | 5/8" rigid insulation board |
| R-7 | 2 1/2" fiberglass batt |
| R-11 | 3 1/2" fiberglass batt |
| R-19 | 3 5/8" fiberglass batt and 1" polystyrene sheathing, or one 6 1/2" batt |

### Floors

| | |
|---|---|
| R-11 | 3 1/2" fiberglass batt or blanket |
| R-13 | 3 5/8" fiberglass batt or blanket |
| R-19 | 6 1/2" fiberglass batt or blanket |
| R-22 | 7" fiberglass batt or blanket |

# FINANCING

Types of Mortgages
Home Improvement Loan Table
Mortgage Payment Table

# FINANCING

## TYPES OF HOME MORTGAGES

| | |
|---|---|
| **Adjustable-rate (ARM)** | Interest rate tied to published financial index such as prime lending rate. Most have annual and lifetime caps. |
| **Assumable** | Buyer takes over seller's mortgage. |
| **Balloon** | Payments based on long term, but principle due in short term. |
| **Blanket Mortgage** | One mortgage covering several properties. |
| **Buy-down** | Seller pays part of interest for first years. |
| **Construction Mortgage** | Finance construction of improvements. |
| **Fixed-rate (FRM)** | Interest rate and monthly payments constant for life of loan (usually 30 years). |
| **Graduated-payment (GPM)** | Payments increase for first few years, then remain constant. Interest rate may vary. |
| **Growing-equity (GEM)** | Payments increase annually, with increase applied to principle. Interest rate usually constant. |
| **Interest-only** | Entire payment goes toward interest only. Principle remains at original amount. |
| **Leasehold Mortgage** | Usually obtained to construct improvements. |
| **Open-end Mortgage** | Additional funds with same agreement. |
| **Owner-financed (seller take back)** | Seller holds either first or second mortgage. |
| **Purchase Money Mortgage** | Purchase subject property. |
| **Renegotiable (rollover)** | Same as adjustable-rate mortgage, but interest rate adjusted less often. |
| **Reverse annuity** | Lender makes monthly payments to borrower. Debt increases over time to maximum percentage of appraised value of property. |

# FINANCING

## TYPES OF HOME MORTGAGES (Cont'd)

**Shared-appreciation (SAM)**    Lender charges less interest in exchange for share of appreciation when property sold.

**Variable Rate Mortgage**    Interest rate varies with a standard rate.

**Zero-interest (no-interest)**    No interest charged. Fixed monthly payments usually over short term.

# FINANCING

## HOW TO USE THE FOLLOWING TABLES

### Monthly Payment

The following tables show monthly payments for each $1,000, borrowed for fixed interest rates, from 7 to 20 3/4 percent and for periods of 5 to 30 years.

### Example:

What are the monthly payments on a 10-year, $20,000 loan at 12% interest? From the table, the monthly payment for each $1,000 is $14.35. Therefore the total monthly payment is 20 x $14.35 = $287.00.

### Total Interest Paid

Knowing the monthly payment, it's easy to find the total interest paid over the life of a loan. Total interest is the difference between the sum of the payments over life and the original amount.

### Example:

For a 25-year, 10 1/2 percent, $100,000 mortgage, what is the total interest paid? From the table, the monthly payment for each $1,000 is $9.45. Therefore, the total lifetime payment is 100 x 12 months x 25 years x $9.45 = $283,500. Since the original loan is for $100,000, the total interest paid is the difference: $183,500.

# FINANCING

## HOME IMPROVEMENT LOAN AND MORTGAGE PAYMENT TABLES
### Monthly Payment, Fixed Term & Interest per $1,000

| TERM<br>RATE | 5 | 10 | 15 | 20 | 25 | 30 |
|---|---|---|---|---|---|---|
| 7 % | 19.81 | 11.62 | 8.99 | 7.76 | 7.07 | 6.66 |
| 7 1/4 | 19.92 | 11.75 | 9.13 | 7.91 | 7.23 | 6.83 |
| 7 1/2 | 20.04 | 11.88 | 9.28 | 8.06 | 7.39 | 7.00 |
| 7 3/4 | 20.16 | 12.01 | 9.42 | 8.21 | 7.56 | 7.17 |
| 8 | 20.28 | 12.14 | 9.56 | 8.37 | 7.72 | 7.34 |
| 8 1/4 | 20.40 | 12.27 | 9.71 | 8.53 | 7.89 | 7.52 |
| 8 1/2 | 20.52 | 12.40 | 9.85 | 8.68 | 8.06 | 7.69 |
| 8 3/4 | 20.64 | 12.54 | 10.00 | 8.84 | 8.23 | 7.87 |
| 9 | 20.76 | 12.67 | 10.15 | 9.00 | 8.40 | 8.05 |
| 9 1/4 | 20.88 | 12.81 | 10.30 | 9.16 | 8.57 | 8.23 |
| 9 1/2 | 21.01 | 12.94 | 10.45 | 9.33 | 8.74 | 8.41 |
| 9 3/4 | 21.13 | 13.08 | 10.60 | 9.49 | 8.92 | 8.60 |
| 10 | 21.25 | 13.22 | 10.75 | 9.66 | 9.09 | 8.78 |
| 10 1/4 | 21.38 | 13.36 | 10.90 | 9.82 | 9.27 | 8.97 |
| 10 1/2 | 21.50 | 13.50 | 11.06 | 9.99 | 9.45 | 9.15 |
| 10 3/4 | 21.62 | 13.64 | 11.21 | 10.16 | 9.63 | 9.34 |
| 11 | 21.75 | 13.78 | 11.37 | 10.33 | 9.81 | 9.53 |
| 11 1/4 | 21.87 | 13.92 | 11.53 | 10.50 | 9.99 | 9.72 |
| 11 1/2 | 22.00 | 14.06 | 11.69 | 10.67 | 10.17 | 9.91 |
| 11 3/4 | 22.12 | 14.21 | 11.85 | 10.84 | 10.35 | 10.10 |
| 12 | 22.25 | 14.35 | 12.01 | 11.02 | 10.54 | 10.29 |
| 12 1/4 | 22.38 | 14.50 | 12.17 | 11.19 | 10.72 | 10.48 |
| 12 1/2 | 22.50 | 14.64 | 12.33 | 11.37 | 10.91 | 10.68 |
| 12 3/4 | 22.63 | 14.79 | 12.49 | 11.54 | 11.10 | 10.87 |
| 13 | 22.76 | 14.94 | 12.66 | 11.72 | 11.28 | 11.07 |
| 13 1/4 | 22.89 | 15.08 | 12.82 | 11.90 | 11.47 | 11.26 |
| 13 1/2 | 23.01 | 15.23 | 12.99 | 12.08 | 11.66 | 11.46 |
| 13 3/4 | 23.14 | 15.38 | 13.15 | 12.26 | 11.85 | 11.66 |
| 14 | 23.27 | 15.53 | 13.32 | 12.44 | 12.04 | 11.85 |
| 14 1/4 | 23.40 | 15.68 | 13.49 | 12.62 | 12.23 | 12.05 |
| 14 1/2 | 23.53 | 15.83 | 13.66 | 12.80 | 12.43 | 12.25 |
| 14 3/4 | 23.66 | 15.99 | 13.83 | 12.99 | 12.62 | 12.45 |
| 15 | 23.79 | 16.14 | 14.00 | 13.17 | 12.81 | 12.65 |
| 15 1/4 | 23.93 | 16.29 | 14.17 | 13.36 | 13.01 | 12.85 |

## FINANCING

## HOME IMPROVEMENT LOAN AND
## MORTGAGE PAYMENT TABLES (Cont'd)
Monthly Payment, Fixed Term & Interest per $1,000

| TERM RATE | 5 | 10 | 15 | 20 | 25 | 30 |
|---|---|---|---|---|---|---|
| 15 1/2 | 24.06 | 16.45 | 14.34 | 13.54 | 13.20 | 13.05 |
| 15 3/4 | 24.19 | 16.60 | 14.52 | 13.73 | 13.40 | 13.25 |
| 16 | 24.32 | 16.76 | 14.69 | 13.92 | 13.59 | 13.45 |
| 16 1/4 | 24.46 | 16.91 | 14.87 | 14.11 | 13.79 | 13.65 |
| 16 1/2 | 24.59 | 17.07 | 15.04 | 14.29 | 13.99 | 13.86 |
| 16 3/4 | 24.72 | 17.23 | 15.22 | 14.48 | 14.18 | 14.06 |
| 17 | 24.86 | 17.38 | 15.40 | 14.67 | 14.38 | 14.26 |
| 17 1/4 | 24.99 | 17.54 | 15.57 | 14.86 | 14.58 | 14.46 |
| 17 1/2 | 25.13 | 17.70 | 15.75 | 15.05 | 14.78 | 14.67 |
| 17 3/4 | 25.26 | 17.86 | 15.93 | 15.25 | 14.98 | 14.87 |
| 18 | 25.40 | 18.02 | 16.11 | 15.44 | 15.18 | 15.08 |
| 18 1/4 | 25.53 | 18.18 | 16.29 | 15.63 | 15.38 | 15.28 |
| 18 1/2 | 25.67 | 18.35 | 16.47 | 15.82 | 15.58 | 15.48 |
| 18 3/4 | 25.81 | 18.51 | 16.65 | 16.02 | 15.78 | 15.69 |
| 19 | 25.95 | 18.67 | 16.83 | 16.21 | 15.98 | 15.89 |
| 19 1/4 | 26.08 | 18.84 | 17.02 | 16.41 | 16.18 | 16.10 |
| 19 1/2 | 26.22 | 19.00 | 17.20 | 16.60 | 16.39 | 16.30 |
| 19 3/4 | 26.36 | 19.17 | 17.38 | 16.80 | 16.59 | 16.51 |
| 20 | 26.50 | 19.33 | 17.57 | 16.99 | 16.79 | 16.72 |
| 20 1/4 | 26.64 | 19.50 | 17.75 | 17.19 | 16.99 | 16.92 |
| 20 1/2 | 26.78 | 19.66 | 17.94 | 17.39 | 17.20 | 17.13 |
| 20 3/4 | 26.92 | 19.83 | 18.12 | 17.58 | 17.40 | 17.33 |

# DEPRECIATION

Types of Depreciation
Definitions & Examples
Building Depreciation Table

# DEPRECIATION

Depreciation is a loss in value due to any cause. It's the difference between the value of a building or equipment and its reproduction or replacement cost new as of the date of valuation. There are three generally accepted causes of depreciation:

## PHYSICAL DEPRECIATION, FUNCTIONAL OBSOLESCENCE AND EXTERNAL OBSOLESCENCE

**PHYSICAL DEPRECIATION** is a loss in value due to physical deterioration. There are two types which are included in the depreciation tables that follow.

> **Curable Physical Depreciation** is generally associated with short-lived items such as paint, floor covers, roof cover and hot-water heaters.
>
> **Incurable Physical Depreciation** is generally associated with long-lived items such as floor structures, roof structures, mechanical systems and foundations. Such items are not normally replaced in a typical maintenance program and are usually incurable except through major reconstruction.

Condition is a measure of the degree at which items subject to physical depreciation have been maintained. The observed condition of each component subject to wear is estimated relative to new condition. Many portions of a structure wear out slowly, if at all, such as excavations, foundations and concrete exterior walls. Such long-lived portions could represent most of the total reproduction cost and, if still functional, will contribute toward an extended life expectancy. Physical depreciation can't be considered a straight-line deduction from reproduction cost, since necessary and normal maintenance can offset, retard and even eliminate deterioration.

**FUNCTIONAL OBSOLESCENCE** is the loss in value due to lack of utility or desirability of all or part of the property. It's inherent to the improvement, included in the tables and may be curable or incurable. A new structure may suffer from functional obsolescence.

# DEPRECIATION

**Curable Functional Obsolescence** is the lack of utility or desirability due to such factors as outdated or inadequate lighting and plumbing fixtures or heating units that could be corrected or replaced without major rebuilding.

**Incurable Functional Obsolescence** is the lack of utility or desirability due to such factors as poorly designed floor plans that interfere with proper space utilization (i.e., many small rooms, tandem rooms, etc.) or antiquated mechanical systems that may be completely incurable except by major reconstruction or renovation and usually are not economically feasible to correct.

**EXTERNAL OBSOLESCENCE** is a loss in value due to causes outside the property and, therefore, is not considered in the following depreciation guidelines. External obsolescence must be determined from studies of the neighborhood or the community as a whole.

## EXPLANATION OF DEPRECIATION TABLES

The depreciation concept in this section is based on an extended life theory which encompasses a remaining life and effective age approach.

**EXTENDED LIFE EXPECTANCY** is the increased life expectancy due to seasoning and proven ability to exist. Just as a person will have a total normal life expectancy at birth which increases as he grows older, so it is with structures and equipment.

**REMAINING LIFE** is the normal remaining life expectation. It's the length of time the structure may be expected to continue to perform its function economically. Determination of effective age on older structures may best be calculated by establishing a remaining life which, when subtracted from a typical life expectancy, will result in an appropriate effective age with which to work.

# DEPRECIATION

**EFFECTIVE AGE** is the age of a structure as compared to other structures performing like functions. It's the chronological age less any age that has been taken off due to face-lifting, remodeling, structural reconstruction, correction of functional inadequacies, modernization of equipment, etc. It's an age that reflects a true remaining life for the property, taking into account the typical life expectancy of buildings or equipment of its type and usage. The effective age is a matter of judgment, taking all factors into consideration.

The depreciation tables were developed from actual case studies of sales and market value appraisals. From confirmed sales prices, the land value was deducted to obtain a building residual and the replacement cost of the building was computed. The difference between the replacement cost new of the building and the residual sales price of the building was divided by the replacement cost new to give the market depreciation in percentage. A similar procedure was followed with the market value appraisals, always excluding those cases having excessive economic obsolescence.

The data was then collated by type of construction and usage, plotted with similar typical total life expectancies, with curves computed for the groupings for which sufficient data was available for statistical reliability. From these curves, a matching family of mathematical curves was found from which the depreciation for any initial life expectancy could be computed.

# DEPRECIATION

## TYPICAL BUILDING LIVES

Typical life expectancies of single and multi-family residences are based on case studies of both actual mortality and ages at which major reconstruction has taken place. The exceptions to the studies are typical life expectancies for modular structures. Typical life expectancies for modular structures assume conformity to site-built residences in both quality and design. All cases of abnormal or excessive obsolescence due to external causes outside of and not inherent to the subject properties were excluded.

| QUALITY | SINGLE-FAMILY (Detached) Site-built or modular: Frame/Masonry | MULTIPLE-FAMILY and SINGLE-FAMILY (Attached) Site-built or modular: Frame/Masonry |
|---|---|---|
| Low | 40/45 | ------ |
| Fair | 45/50 | 40/45 |
| Average | 50/55 | 45/50 |
| Good | 50/55 | 45/50 |
| Very Good | 55/60 | 50/55 |
| Excellent | 55/60 | ------ |

**Use of the tables:**
1. Determine the chronological age of the residence.
2. Compare the subject residence with like properties and study the effect of any modernization or major repair to determine effective age.
3. Determine typical life expectancy from table above.
4. Enter the depreciation table in the column for the appropriate life expectancy and at the effective age estimated in Step 2. The corresponding number is a normal percentage of depreciation.

# DEPRECIATION

## Condition Modifier Table

The Depreciation Table is based on normal maintenance for structures of like type, age and occupancy. An adjustment may be necessary for above-average or deferred maintenance if you choose to treat condition separately. The effective age of a newer residence is usually the same as its chronological age and will not require a condition modification until those items subject to physical depreciation begin to show wear and may need repair.

**EXCELLENT CONDITION** – All items that can normally be repaired or refinished have recently been corrected, such as new roofing, new paint, furnace overhaul, etc . . . . . . . . . . . . . . . . . . . . . . . . . . . . . . . . . . . . . . . x  .80

**VERY GOOD CONDITION** – All items well maintained, many having been overhauled and repaired as they showed signs of wear . . . . . x  .85

**GOOD CONDITION** – No obvious maintenance required but neither is everything new  . . . . . . . . . . . . . . . . . . . . . . . . . . . . . . . . . . . . . . . . . x  .90

**AVERAGE CONDITION** – Some evidence of deferred maintenance in that a few minor repairs are needed along with some refinishing  . . . . x 1.00

**BADLY WORN** – Much repair needed. Many items need refinishing or overhauling  . . . . . . . . . . . . . . . . . . . . . . . . . . . . . . . . . . . . . . . . . . x 1.10

**WORN OUT** – Repair and overhaul needed on painted surfaces, roofing, plumbing, heating, etc. (Found only in extraordinary circumstances)
. . . . . . . . . . . . . . . . . . . . . . . . . . . . . . . . . . . . . . . . . . . . . . . . . . . . . x 1.15

## EXAMPLE

The subject residence is a good-quality, wood frame, single-family, detached residence. From inspection of the residence and through comparison to other similar residences, the appraiser has determined that the effective age and the chronological age for the subject and the neighborhood are equal and are 20 years.

# DEPRECIATION

The typical life expectancy from the table on the previous page is 50 years. From the depreciation table, the percentage of depreciation is 22%. Since the subject residence was in somewhat better condition than average for the area, a condition modifier for good condition at .90 is used in place of lowering the overall effective age. This modifier is multiplied by the percentage of depreciation to estimate a net depreciation of 19.8% (.90 x 22% = 19.8%). Rounded, this represents a 20% depreciation at a 50-year life.

Referring again to the depreciation table to compare the modified effective age after applying a condition modifier, the net depreciation of 20% is that of a residence having an effective age of 19 years instead of a 20-year effective age that was typical of the neighborhood.

# DEPRECIATION

## TYPICAL LIFE EXPECTANCY IN YEARS

| Effective Age In Years | 15 | 20 | 25 | 30 | 35 | 40 | 45 | 50 | 55 | 60 | Effective Age In Years |
|---|---|---|---|---|---|---|---|---|---|---|---|
| | | | | **Depreciation – Percentage** | | | | | | | |
| 1  | 5%  | 4%  | 3%  | 3%  | 3%  | 2%  | 1%  | 1%  | 0%  | 0%  | 1  |
| 2  | 9   | 7   | 6   | 6   | 5   | 4   | 2   | 1   | 1   | 1   | 2  |
| 3  | 14  | 11  | 8   | 8   | 8   | 6   | 3   | 2   | 2   | 2   | 3  |
| 4  | 18  | 14  | 11  | 10  | 10  | 7   | 4   | 3   | 2   | 2   | 4  |
| 5  | 22  | 17  | 13  | 12  | 12  | 8   | 5   | 4   | 3   | 3   | 5  |
| 6  | 26  | 20  | 16  | 14  | 14  | 10  | 6   | 5   | 4   | 4   | 6  |
| 7  | 30  | 23  | 18  | 16  | 15  | 12  | 7   | 6   | 5   | 5   | 7  |
| 8  | 33  | 26  | 21  | 19  | 17  | 13  | 8   | 7   | 6   | 6   | 8  |
| 9  | 37  | 29  | 23  | 21  | 19  | 15  | 9   | 8   | 7   | 6   | 9  |
| 10 | 41  | 32  | 25  | 23  | 21  | 16  | 11  | 9   | 8   | 7   | 10 |
| 11 | 45  | 35  | 27  | 25  | 23  | 18  | 12  | 10  | 9   | 8   | 11 |
| 12 | 48  | 37  | 29  | 26  | 24  | 19  | 13  | 11  | 10  | 9   | 12 |
| 13 | 52  | 40  | 32  | 28  | 26  | 21  | 14  | 12  | 11  | 10  | 13 |
| 14 | 55  | 43  | 34  | 30  | 28  | 23  | 16  | 13  | 12  | 11  | 14 |
| 15 | 59  | 46  | 37  | 32  | 30  | 24  | 17  | 15  | 13  | 12  | 15 |
| 16 | 63  | 48  | 39  | 34  | 31  | 26  | 19  | 16  | 14  | 13  | 16 |
| 17 | 66  | 51  | 41  | 36  | 33  | 28  | 20  | 17  | 16  | 14  | 17 |
| 18 |     | 54  | 44  | 38  | 34  | 29  | 22  | 19  | 17  | 16  | 18 |
| 19 |     | 57  | 46  | 40  | 35  | 30  | 24  | 20  | 18  | 17  | 19 |
| 20 |     | 60  | 48  | 41  | 37  | 31  | 25  | 22  | 20  | 18  | 20 |
| 21 |     | 63  | 50  | 43  | 39  | 33  | 26  | 24  | 21  | 19  | 21 |
| 22 |     | 66  | 52  | 45  | 41  | 35  | 28  | 25  | 22  | 20  | 22 |
| 23 |     |     | 55  | 47  | 42  | 36  | 30  | 26  | 24  | 21  | 23 |
| 24 |     |     | 57  | 49  | 44  | 37  | 32  | 28  | 25  | 23  | 24 |
| 25 |     |     | 61  | 51  | 46  | 39  | 33  | 30  | 27  | 24  | 25 |
| 26 |     |     | 63  | 52  | 47  | 40  | 35  | 32  | 28  | 25  | 26 |
| 27 |     |     | 65  | 54  | 49  | 42  | 36  | 33  | 29  | 27  | 27 |
| 28 |     |     |     | 56  | 50  | 43  | 38  | 35  | 31  | 28  | 28 |
| 29 |     |     |     | 58  | 52  | 45  | 40  | 36  | 33  | 29  | 29 |
| 30 |     |     |     | 60  | 54  | 46  | 41  | 38  | 34  | 31  | 30 |
| 31 |     |     |     | 62  | 55  | 48  | 42  | 40  | 36  | 32  | 31 |
| 32 |     |     |     | 63  | 56  | 49  | 44  | 42  | 38  | 33  | 32 |
| 33 |     |     |     | 65  | 58  | 51  | 46  | 44  | 39  | 35  | 33 |

# DEPRECIATION

## TYPICAL LIFE EXPECTANCY IN YEARS (Cont'd)

| Effective Age In Years | 15 | 20 | 25 | 30 | 35 | 40 | 45 | 50 | 55 | 60 | Effective Age In Years |
|---|---|---|---|---|---|---|---|---|---|---|---|
|   |   |   |   |   | **Depreciation – Percentage** |   |   |   |   |   |   |
| 34 |   |   |   |   | 59 | 52 | 48 | 45 | 41 | 36 | 34 |
| 35 |   |   |   |   | 61 | 53 | 49 | 46 | 42 | 37 | 35 |
| 36 |   |   |   |   | 62 | 54 | 51 | 48 | 43 | 39 | 36 |
| 37 |   |   |   |   | 64 | 56 | 53 | 49 | 45 | 40 | 37 |
| 38 |   |   |   |   | 66 | 57 | 54 | 51 | 46 | 41 | 38 |
| 39 |   |   |   |   |    | 59 | 55 | 52 | 47 | 41 | 39 |
| 40 |   |   |   |   |    | 60 | 56 | 53 | 49 | 43 | 40 |
| 41 |   |   |   |   |    | 62 | 57 | 54 | 50 | 45 | 41 |
| 42 |   |   |   |   |    | 63 | 58 | 56 | 51 | 46 | 42 |
| 43 |   |   |   |   |    | 63 | 59 | 57 | 52 | 47 | 43 |
| 44 |   |   |   |   |    | 64 | 60 | 58 | 54 | 48 | 44 |
| 45 |   |   |   |   |    | 65 | 61 | 58 | 55 | 49 | 45 |
| 46 |   |   |   |   |    |    | 61 | 59 | 56 | 50 | 46 |
| 47 |   |   |   |   |    |    | 62 | 60 | 57 | 51 | 47 |
| 48 |   |   |   |   |    |    | 63 | 61 | 58 | 52 | 48 |
| 49 |   |   |   |   |    |    | 64 | 61 | 58 | 53 | 49 |
| 50 |   |   |   |   |    |    | 65 | 62 | 59 | 54 | 50 |
| 51 |   |   |   |   |    |    |    | 62 | 60 | 55 | 51 |
| 52 |   |   |   |   |    |    |    | 63 | 61 | 56 | 52 |
| 53 |   |   |   |   |    |    |    | 64 | 61 | 57 | 53 |
| 54 |   |   |   |   |    |    |    | 64 | 62 | 58 | 54 |
| 55 |   |   |   |   |    |    |    |    | 62 | 59 | 55 |
| 56 |   |   |   |   |    |    |    |    | 62 | 59 | 56 |
| 57 |   |   |   |   |    |    |    |    | 63 | 60 | 57 |
| 58 |   |   |   |   |    |    |    |    | 63 | 61 | 58 |
| 59 |   |   |   |   |    |    |    |    | 64 | 62 | 59 |
| 60 |   |   |   |   |    |    |    |    | 64 | 62 | 60 |
| 61 |   |   |   |   |    |    |    |    | 65 | 63 | 61 |
| 62 |   |   |   |   |    |    |    |    |    | 63 | 62 |
| 63 |   |   |   |   |    |    |    |    |    | 64 | 63 |
| 64 |   |   |   |   |    |    |    |    |    | 64 | 64 |
| 65 |   |   |   |   |    |    |    |    |    | 64 | 65 |

# GLOSSARY/
INDEX

# GLOSSARY

**Explanation**

The purpose of the following definitions is to provide a better understanding of key terms. It's not the intention of this section to serve as a comprehensive appraisal, architectural or construction dictionary.

## DEFINITIONS:

**Abutment**  A foundation structure designed to withstand thrust, such as the end supports of an arch.

**Acoustical ceiling**  In general terms, a ceiling designed to lessen sound reverberation by absorption, blocking or muffling. In construction, the most common materials are acoustical tile and acoustical plaster.

**Adobe**  Solid masonry wall made from adobe block, which is unburnt, sun-dried blocks molded from adobe soil found in arid regions, generally rough in shape and texture. The wall may be grouted and reinforced or of a post and girder-type construction. Modern adobe can have an asphalt or chemical binder.

**A-frame**  The structural support framework in the shape of the letter "A". Also a building system having sloping side members which act as both walls and roof.

**Aggregate**  All the materials used in the manufacture of concrete or plaster except water and the bonding agents (cement, lime, plaster). May include sand, gravel, cinders, rock, slag, etc.

**Air conditioning**  The process of bringing air to a required state of temperature and humidity, and removing dust, pollen or other foreign matter.

**Air infiltration wrap**  A high-density polyethylene fibrous exterior air barrier generally applied to residential stud construction.

**Appliance allowance**  This cost includes consideration for the residential appliances commonly found at different quality levels. Typically, ranges and ovens, garbage disposals, dishwashers and range hoods are included. The better qualities (higher cost ranks) have additional feature considerations for trash compactors, microwaves, built-in mixer units, etc.

**Areaway**  Any area that is open and below grade which allows access to a basement or crawl space and provides light or air into that area.

**Areaway wall**  A wall which surrounds an open area such as an areaway.

# GLOSSARY

**Ash dump** An area under a fireplace which allows for the removal of ashes which have been dumped from the fireplace above.

**Asphalt shingles** A type of shingle made of felt, saturated with asphalt or tar pitch and surfaced with mineral granules or inorganic fiberglass saturated with asphalt and surfaced with ceramic granules. There are many different patterns, some individual and others in strips.

**Asphalt tile** A resilient floor covering laid in mastic which is available in a variety of colors. Standard size is 9" x 9". Asphalt is normally used only in the darker colors with the lighter colors generally having a resin base.

**Atrium** An interior courtyard usually with a glass roof to provide a greenhouse–like effect inside.

**Attic** A room built within the sloping roof of a dwelling. May be finished or unfinished.

**Backfill** Material used in refilling an excavation, such as for a foundation or subterranean pipe.

**Backup** A lower-priced material in a masonry wall that is covered by a facing of a more expensive and ornamental material, such as face brick, stone or marble.

**Balcony** A railed platform projecting from the face of a building above the ground level, with an entrance from the building interior.

**Balusters** The closely spaced vertical members in a stairway or balcony, balustrade or railing.

**Baseboard** Often referred to as a base or mopboard, this board or molding is installed against the wall where the wall and floor meet. It can be found in any number of sizes and shapes.

**Baseboard heating** Heating in which the radiant heating element, usually an electric resistance unit or forced hot water, is located at the base of an interior wall.

**Basement** Any room or rooms built partially or wholly below ground level.

**Batt insulation** A type of blanket insulating material, usually composed of mineral fibers and made in relatively narrow widths for convenience in handling and application between framing members.

# GLOSSARY

**Batten**  A narrow strip of wood used to cover a joint between boards, or to simulate a covered joint for architectural purposes.

**Bay window**  A window structure that projects from a wall. Technically, it has its own foundation. If cantilevered, it would be an oriel window, however, in common usage, the terms are often used interchangeably

**Beam**  A horizontal, load-bearing structural member, transmitting vertical loads to walls, columns or heavier horizontal members.

**Beamed ceiling**  A ceiling with beams exposed. A false-beamed ceiling has ornamental boards or timbers which are not load bearing.

**Bearing wall**  A wall that supports upper floor or roof loads.

**Bevel**  A surface cut at other than a right angle.

**Bevel siding**  An exterior wood siding applied horizontally in overlapping rows so that the thick butt overlaps the thin edge of the board below. Also referred to as clapboard.

**Bi-level**  A two-story residence with a split-foyer entrance. The lower level, partially above grade, is partially finished. Typically, the finish includes plumbing and electrical rough-ins with some partition wall framing for recreation room, bedroom, laundry area and bathroom. Other common terms for this type of construction are raised ranch, hi-ranch, colonial and split entry.

**Black top**  A general term to describe asphalt or asphaltic concrete paving.

**Blanket insulation**  A flexible type of lightweight blanket for insulating purposes. It's supplied in rolls, strips or panels, sometimes fastened to heavy paper of an asphalt-treated or vapor-barrier type. Blankets may be composed of various processed materials, such as mineral wool, wood, glass or glass fibers.

**Blind**  When inside a house, a window covering which can be raised or lowered to provide privacy or block out light.

**Brick veneer**  A non-loadbearing single tier of brick applied to a wall or other materials.

**Brick veneer wall**  Usually used to describe a wall made up of brick veneer applied over wood framing.

**Bridging**  Diagonal or cross bracing between joists to resist twisting.

# GLOSSARY

**B.T.U.**  British thermal unit. A measurement of heat, (i.e., the amount of heat required to raise one pound of water by one degree Fahrenheit).

**Brownstone**  A term usually referring to houses built, until about 1900, with a brown-colored, quarried, thick-cut solid sandstone which was laid up in mortar.

**Building paper**  A paper usually applied over the sheathing of exterior frame walls. Also used between flooring and subflooring and over roof decks.

**Building permit**  A certificate that must be obtained from the municipal government by the property owner or contractor before a building can be erected or repaired. It must be kept posted in a conspicuous place until the job is completed and passed by the building inspector.

**Built-in appliances**  Those appliances that are permanent fixtures generally found in the residence. They are not included in the base costs and should be added separately.

**Cantilever**  A beam or slab supported at one end only, or which projects beyond its support.

**Caissons**  Poured-in-place, reinforced concrete pilings.

**Carport**  An open automobile shelter. May be only a roof and supports or may be enclosed on three sides with one completely open side.

**Casement window**  A window hinged vertically, swinging open horizontally like a door.

**Caulking**  Material used to seal cracks, fill joints, and prevent seepage. Includes mastic compounds with silicone, asphalt or rubber bases.

**Ceiling joist**  The structural members to which the ceiling is fastened.

**Cellar**  An enclosd area that is generally used for storage which is partially or completely underground.

**Cement fiber (asbestos) shingles**  A covering, consisting largely of Portland Cement and asbestos fiber, made into shingles.

**Cesspool**  A pit that stores liquid sewage, which is disposed of through seepage into the surrounding soil.

**Chair rail**  A horizontal piece of wood which runs along the wall at a height intended to protect the wall from chairs scratching the wall at that level (about 3' from the floor).

# GLOSSARY

**Chimney**  A vertical structure which is made up of individual flues which is enclosed within a masonry support work.

**Chimney cap**  The masonry top piece of a chimney or a metal piece that tops the flues of a chimney to improve the draft while minimizing the entry of rain into the chimney.

**Chimney flashing**  Strips of metal sheet or other materials which are attached where the base of the chimney meets the roof which are used to waterproof the joints.

**Chimney stack**  A vertical vent designed to dispose of waste gases and heat and to create a shaft for furnaces or boilers.

**Clerestory window**  A series or band of vertical windows set above the primary roof line.

**Collar beam**  A horizontal beam which attaches to two opposing rafters approximately at midpoint to improve stability.

**Column**  A vertical structural member; a pillar. False columns are designed for architectural ornamentation rather than load-bearing qualities.

**Common wall**  A single wall used jointly by two buildings, also called a party wall.

**Composition shingle**  A roofing shingle made of either felt saturated with asphalt and surfaced with mineral granules or inorganic fiberglass saturated with asphalt and surfaced with ceramic granules.

**Concrete basement floor**  A basement floor made of a composite material consisting of a coarse aggregate, fine aggregate, cement and water.

**Condominium**  Type of ownership of a multi-unit property in which the owner holds title to an individual unit and a percentage of common areas.

**Conduit**  A pipe or channel carrying electrical wiring, water or other fluids. It may be rigid or flexible.

**Corner brace**  A diagonal brace at the corner of a wood-frame building used to reinforce the building.

**Corner stud**  A vertical post which is found at the corner of a wood structure.

**Cornice moulding**  An ornamental trim found on the exterior of a house which is placed where the roof and wall meet.

# GLOSSARY

**Coved ceiling** A ceiling which curves down at the edges where it meets the wall, providing a smooth transition from ceiling to walls instead of a sharp angle of intersection.

**Crawl space** A space of limited height sufficient to permit access to piping or wiring underneath the floor of a raised-floor structure.

**Cripple stud** A stud that is less than full length.

**Cupola** A small square or rectangular structure located along the roof ridge used for ventilation and/or ornamentation.

**Curtain wall** A non-bearing exterior wall supported by an independent structural frame of a building.

**Detached dwelling** A housing unit or garage with wall and roof independent of any other building, as opposed to an attached dwelling.

**Diagonal subflooring** Subflooring which has been laid on the diagonal to the floor joist.

**Doorjamb** The vertical piece on either side of a door frame as well as the horizontal piece atop the door.

**Door trim** The molding around a door frame that covers the joint between the frame and the wall.

**Dormer** A projection from a sloping roof to provide more headroom under the roof and allow the installation of dormer windows.

**Double glazing** A double-glass pane in a door or window, with an air space between the two panes, which may be sealed hermetically to provide insulation.

**Double-hung window** A window with an upper and lower sash, each balanced by springs or weights enabling vertical movement in its own grooves.

**Downspout or leader gooseneck** A vertical drain which runs from the roof line to the ground and allows water to drain.

**Downspout or leader strap** A piece which secures the downspout to the side of a building.

**Drywall** A finish material applied to an interior wall in a dry state, as opposed to plaster. Normally referred to as gypsum board or sheet rock.

# GLOSSARY

**Ducts**  Enclosures, usually round or rectangular in shape, for distributing warm or cool air from the central unit to the various rooms.

**Eaves**  The portion of a roof projecting beyond the wall line.

**Eave trough or gutter**  A long, shallow trough which runs under the eaves of the roof into which water collects from the roof.

**Electric cable heating**  A heating system consisting of electrical coils installed beneath the surface of ceilings, walls or floors. It is commonly installed in ceilings of multi-family residences having a sprayed-on ceiling.

**Elevated slab**  A horizontal, reinforced, concrete structure that is formed and poured in place above the ground level.

**Elevation**  A scale drawing of the front, rear or side of a building.

**Entrance canopy**  An overhanging roof which protects the entrance of a structure.

**Entrance platform**  The floor or surface which is raised above the ground and provides the entryway into a structure.

**Entrance post**  A vertical support which holds up the entrance canopy.

**Evaporative cooler**  An air conditioner that cools the air by water evaporation, also referred to as a swamp cooler. Outdoor air is drawn through a moistened filter pad in a cabinet, and the cooled air is then circulated throughout the house. This type of cooler is used in regions with low humidity.

**Face brick**  A clay brick made especially for exterior use with special consideration of color, texture and uniformity.

**Fascia**  A horizontal band of material applied at the top of the wall or the end of the eaves as ornamentation and/or to cover the rafter ends.

**Fenestration**  Generally referred to as the arrangement of windows and doors in the walls of a building.

**Finish floor**  The top surface of a floor.

**Finish hardware**  All exposed hardware in a house such as door knobs, door hinges, locks and clothes hooks, etc.

**Fireplace hearth**  The floor of a fireplace and the area immediately adjacent to the fireplace which is made of non-combustible material. The hearth can be flush to the floor or raised above the floor level.

# GLOSSARY

**Floor area**  An area on any floor, enclosed by exterior walls and/or partitions. Measurement for total floor area should include the width of the exterior walls.

**Floor joists**  The joists that support the floor.

**Footing**  The projecting base of a foundation which transmits the building load to the ground.

**Forced air heating**  A warm-air heating system that circulates air by a motor-driven fan. It includes air-cleaning devices.

**Formica**  A trade name for a hard, laminated plastic surfacing, often the name for all such finishes used on counter tops.

**Foundation drain tile**  Perforated piping which collects water below the surface and redirects it away from the structure.

**Foundation wall**  That part of the foundation which forms a retaining wall for that part of the structure which is below ground.

**Frieze**  A horizontal trim piece which is attached at the angle formed between the soffit and the top of the exterior wall covering.

**French doors or windows**  A pair of hinged, glazed doors, functioning as both doors and windows.

**Fresco**  Watercolor painting on damp plaster.

**Furring strips**  Strips of wood or metal which are attached to surfaces to provide a place on which to fasten another surface or to nail.

**Gable roof**  A ridged roof that slopes up from only two walls. A gable is the triangular portion of the end of the building, from the eaves to the ridge.

**Gable stud**  A stud which provides a framework in the gable portion of a structure, which extends from the eaves to the peak of the roof.

**Gambrel roof**  A type of roof that has its slope broken by an obtuse angle, so that the lower slope is steeper than the upper slope; a roof with two pitches.

**Garage cornice**  The exterior ornamental trim which is applied where the roof and garage wall meet.

**Garage door**  The exterior door which covers the opening to a garage area. This door is frequently an electric door which can be opened with a remote control device.

# GLOSSARY

**General contractor**  A builder who is responsible for all work in building a structure.

**Girder**  A horizontal piece which supports the floor joists or subfloor which, in residential construction, is generally made of wood but can also be made of reinforced concrete or steel.

**Girder post**  A support member for a girder.

**Glass block**  A hollow structural glass block laid as masonry for translucent effect in wall construction.

**Grade line**  A line which is staked out before construction begins which serves as the existing or proposed ground level or elevation on a building site.

**Gravity heating**  A warm air system usually located in a basement which operates on the principle of warm air rising through ducts to the upper levels. Since it does not contain a fan as does the conventional forced air furnace, a large burner surface as well as larger ducts are used.

**Grout**  A thin concrete mixture used to fill various voids in masonry work or in other work which requires a very fluid mixture.

**Hardboard**  A highly compressed wood fiberboard with many uses as exterior siding, interior wall covering or concrete forms.

**Header**  In brick masonry construction, a course of brick in which the masonry units are laid perpendicular to the face of the wall to tie two wythes of brick together. In carpentry, a beam carrying a load over an opening — a lintel.

**Heat pump**  This is a self-contained, reverse cycle, heating and cooling unit. On its cooling cycle, it works like an air conditioner, collecting heat from inside and pumping to an outside coil where it's dissipated. On the heating cycle, heat is collected by the outside coil and pumped inside.

**Hipped roof**  A pitched roof having sloping ends rather than gable ends.

**Hot water heating**  The circulation of hot water from a boiler through a system of pipes and radiators or convectors, either by gravity or a circulating pump, allowing the heat to radiate into a room.

**Humidifier**  A device for maintaining desirable humidity conditions in the air supplied to a building.

**Insulation**  Any material used to obstruct the passage of sound, heat, vibration or electricity from one place to another.

# GLOSSARY

**Interim money, cost of**  Interest on financing during a normal period of construction, as well as an amount for servicing or handling of the loan. Bonuses (points) or discounts paid for securing the financing are not included in the costs.

**Jalousie**  An adjustable glass louver. Also refers to doors or windows containing jalousies.

**Keene's cement**  A hard, water-resistant plaster.

**Lath**  Any material used as a base for plaster including wood lath, gypsum lath, wire and metal lath.

**Lintel**  A horizontal framing member carrying a load over a wall opening (also known as a header).

**Lookout**  A horizontal piece which runs from the outside wall to the lower end of a rafter to support the overhanging portion of a roof.

**Mansard roof**  A roof with two slopes -- the lower slope which is very steep and the upper slope which is almost flat.

**Mantle**  The shelf above a fireplace as well as the ornamental trim that surrounds a fireplace opening.

**Masonry construction**  In building, a type of construction with concrete, concrete block or brick load-bearing exterior walls.

**Mesh**  Heavy steel wire welded together in a grid pattern, used as a reinforcement for concrete work.

**Metal lath**  A type of lath made of metal which is used as the base for plaster. It is generally attached to wall studs.

**Millwork**  Wooden portions of a building that have been pre-built and finished in a shop and brought to the site for installation, such as cabinets, door jambs, molding, trim, etc.

**Modular construction**  Any building construction that is normally preassembled and shipped to the site in units.

**Monolithic**  One piece. Monolithic concrete is poured in a continuous process so there are no separations.

**Mortar**  A pasty mixture of cement, lime, sand and water, used as a bonding agent for brick, stone or other masonry units.

# GLOSSARY

**Mud sill**  Timber laid directly on the ground to form the foundation of a structure.

**Newel post**  The post at the bottom of a stair or the end of a flight of stairs to which the balustrade is anchored. The center pole of a spiral staircase.

**Overhead and profit**  Overhead is a contractor's operating expense, including workmen's compensation, fire and liability insurance, unemployment insurance, equipment, temporary facilities, security, etc., that cannot be prorated to any specific category of the construction. Profit is the compensation accrued for the assumption of risk in constructing the building only. This is not to be confused with a developer's or owner's overhead and profit, which is associated with subdivision planning and administration.

**Panel**  Any flat raised or recessed surface in a door, wall ceiling, etc. Any flat sheet of material used as a construction component.

**Parameter**  Any characteristic of a statistical universe that is measurable. In construction: square foot, cubic yard, board feet, etc., are cost parameters.

**Parapet wall**  The portion of a wall that projects above the roof line.

**Parquet flooring**  Wood blocks or strips laid in decorative patterns.

**Perimeter**  The total length of all exterior bearing walls of a building.

**Pier**  The short, individual concrete or masonry foundation supports for the post and girder underpinning of a raised floor structure.

**Pilaster**  A column, usually formed of the same material, and integral with, but projecting from, a wall.

**Pilings**  Columns extending below ground to bear the loads of a structure when the surface soil cannot. They may extend down to bearing soil or support the load by skin friction. Sheet piling is used to form bulkheads or retaining walls.

**Plaster**  Portland Cement mixed with sand and water to form a mortar-like consistency, used for covering walls and ceilings of a building.

**Plumbing fixtures**  Receptacles that receive and discharge water, liquid or waterborne wastes into a drainage system with which they are connected.

**Plywood**  A construction material formed by cementing several sheets of wood face to face, the grain running at right angles in alternate layers.

# GLOSSARY

**Porch**  A wood or concrete platform, often with a roof covering, found at the entrance.

**Precast concrete**  Concrete structural components that are not formed and poured in place within the structure, but are cast separately either at another location or on site.

**Prime coat**  The first coat of paint, an undercoat, to prepare the surface for finish coats.

**Quantity survey**  A method of cost estimation that considers a detailed count of all materials going into a structure together with the cost of labor to install each unit of material.

**R-Value**  The standard measurement of resistance to heat loss related to a given thickness of insulation required by climatic demands.

**Radiant heating**  A system in which a space is heated by the use of hot water pipe coils or electric resistance wires placed normally in the floor or ceiling, allowing the heat to radiate into a room.

**Rafters**  Structural pieces which make up the framing for the roof of a building and support the roof deck and covering.

**Reinforcing steel**  Steel bars used in concrete construction for giving added strength; such bars are of various sizes and shapes.

**Resilient floor covering**  Floor products characterized by having dense, non-absorbent surfaces, available in sheet or tile form. Among the various types are vinyl asbestos tile, asphalt tile, composition tile and linoleum.

**Ridge**  The horizontal line which serves as the junction of the top edge of a double-pitched roof.

**Ridge board**  A board to which rafters are attached which is situated on the edge at the ridge of a roof frame.

**Riser**  The vertical face between two stair treads.

**Roof boards**  Lumber or plywood which is fastened to the top of the rafters and forms an outer sheathing to which the roofing is applied.

**Roofing**  Any material (such as slate, clay or concrete tile, asphalt shingles, wood shingles, etc.) which is applied to the roof to make it waterproof.

**Rough-in**  Drain and water line hookups for laundry facilities or for future fixture installation.

# GLOSSARY

**Sash**  A single frame of a window which holds the glass.

**Septic tank**  A watertight settling tank in which solid sewage is decomposed by natural bacterial action.

**Shake**  A shingle split (not sawed) from a bolt of wood and used for roofing and siding, or it can refer to a manufactured imitation.

**Shutter**  A movable cover or screen to cover an opening.

**Sidewalk**  The pathway which leads from the street or the driveway to a door of a house. It can be made of a number of materials including concrete, brick, tile, etc.

**Sill**  The lowest horizontal framing member of a structure, resting on the ground or on a foundation. Also, the lowest horizontal member of a window or door casing.

**Skylight**  An opening in a roof, covered with plastic or glass, for light and ventilation.

**Slope**  The ratio of rise to run which expresses the angle of a roof pitch.

**Smoke detector**  A fire detector that indicates the presence of smoke based on a light-obscuring principle using photoelectric cells.

**Stair riser**  The vertical board which runs between the treads of a stairway.

**Stair stringer**  The inclined side of a staircase which supports the stair risers and stair treads.

**Storm door**  An extra outside or additional door for protection against inclement weather. Such a door also lessens the chill of a building's interior, making it easier to heat. It also helps to avoid the effects of wind and rain at the entrance doorway.

**Storm window**  A window placed outside an ordinary window for additional protection against severe weather. Also called a storm sash.

**Stucco**  A coating for exterior walls in which cement is put on in wet layers and, when dry, becomes exceedingly hard and durable.

**Stud**  A vertical framing member, either wood or steel, to which wall finishes are attached. Usually, only lumber of 2" x 4" or less dimensions or its steel equivalent are considered as studs. Also, bolt-like components, either threaded or unthreaded, fixed to structural elements to which other elements may be fastened.

# GLOSSARY

**Sump pump** A suction device, usually operated to remove water or waste which collects at the sump pit or tank.

**Tar paper strip** Pieces of tar paper which are used as a waterproofing barrier between the foundation wall and the ground.

**Termite shield** A shield, generally made of galvanized sheet metal, which is placed on a foundation wall to prevent the passage of termites.

**Terra cotta** Hard-burned, unglazed clay, usually molded into shapes for ornamentation of structural surfaces.

**Terrazzo** A floor surface of marble chips in concrete. After the concrete has hardened, the floor is ground and polished to expose the marble chips. In epoxy terrazzo, the filler material is plastic.

**Thermostat** An instrument, electrically operated, which automatically controls the operation of a heating or cooling device by responding to changes in temperature.

**Tie** Any structural member that acts in tension to hold separated structural components together.

**Tongue and groove (T&G)** Any lumber, such as boards or planks, machined in such a manner that there is a groove on one edge and a corresponding projection on the other.

**U factor** The heat transmission factor of a wall, roof or floor assembly measured in BTUs per square foot per degree Fahrenheit.

**Vapor barrier** Material used to retard the passage of moisture through floors, roofs or exterior walls and thus prevents condensation with them; also called moisture barrier. See waterproofing.

**Veneer** A layer of material applied to another surface for ornamental or protective purposes. Masonry veneer refers to any masonry unit applied over wood frame construction.

**Vinyl composition (asbestos) tile** A resilient floor covering laid in mastic that is available in many colors and textures. Standard size is 12" x 12".

**Wainscoting** A decorative surface which extends approximately one-fourth or one-half up the wall, generally to a height of 3 to 4 feet.

**Wall sheathing** A structural covering of lumber or plywood applied to the outside of wall studs.

# GLOSSARY

**Waterproofing**  Any material designed to stop passage of moisture. Plastic sheets of treated papers and asphalt are used for membranes, while various chemical sealants and asphalt applications are used to seal pores or cracks.

**Weather stripping**  Strips of felt, rubber, metal or other suitable material fixed along the edges of a door or window to keep out drafts and reduce heat loss.

**Window casing**  The exposed trim around windows. There is both an interior and exterior window casing for each window.

**Window cripple**  Studs which are found above or below a window which are less than full length (see cripple studs).

**Window frame**  The framework in which the sash is housed to which it is fixed or hinged.

**Window light**  A pane of glass in a window.

**Wood frame construction**  In building, a type of construction in which the structural members are wood or are dependent upon a wood frame for support. Same as frame construction.

# INDEX

## A

A-frame, 79, 97
abutment, 97
accoustical custom spray, 11
acoustical ceiling, 97
acoustical panel, 11
acoustical panel w/suspension grid, 11
acoustical tile, 11
acoustical tile w/furring, 11
acrylic/plexiglas, 71
adobe, 97
aggregate, 97
air conditioner (window), 42
air conditioning, 97
air duct, 42
air filtration wrap, 97
air intake grille, 42
air purifier (electronic), 42
air purifier (filtered), 42
air register (return), 42
air register (supply), 42
alarm base, security, 25
alarm points, base, 25
alarm station, fire, 24
aluminum or vinyl siding, 31
aluminum storm door, 14

antenna (TV/radio), 23
appliance allowance, 99
appliance hookup, 50
areas, computing, 79
areaway, 97
areaway wall, 97
arrester, lightning, 25
ash dump, 100
asphalt tile, 100
asphalt (hot mopped), 56
asphalt paving, 60
asphalt shingles, 100
asphalt tile, 36
atrium, 100
attic, 100
average story height, determining, 78
awning windows, 71
awnings, 71

## B

B.T.U., 102
backfill, 100
backup, 100
balcony, 100
barbeque (built-in), 60
base cabinet, 46
base station, telephone, 26
baseboard, 100

# INDEX

baseboard heating, 100
basement, 100
balusters, 100
bathroom accessories set, 6
bathroom layout, typical, 5
bathroom, room lighting standards, 19
bathtub, 6
bathtub enclosure, 6
bathtub/shower combination, 6
batt insulation, 100
batten, 101
bay window, 101
beam, 101
beamed ceiling, 101
bearing wall, 101
bedroom, room lighting standards, 19
bevel, 101
bevel siding, 101, 103
bi-level, 101
bib, hose, 50
bidet, 6
bifold, closet, 16
black top, 101
blanket insulation, 101
blender (food center), 46
blind, 101
blinds, 61
block (concrete), 66
block walls, 60
block, masonry, 32
blower (ventilation), 42
board, gypsum, 66
board, insulating, 66
boards, hardboard, 32
boiler (hot water), 42
box, outlet, 25
brick (solid), 32, 66
brick pavers, 36
brick paving, 60
brick veneer, 32, 66, 101
brick veneer (simulated), 66
brick veneer wall, 101
brick wall patterns, 29
brick walls, 60
bridging, 101
broom closet, 61
brownstone, 102
building lives, typical, 94
building paper, 102
building permit, 102
built-in appliances, 102
built-up, 56

# INDEX

## C

cabinet, base, 46
cabinet, wall, 47
cable (wire), 23
caissons, 102
candelabra, 23
canopy, entrance, 105
cantilever, 102
carpeting, 36
carport, 102
casement windows, 71, 102
casing, window, 111
caulking, 32, 66, 102
caulking (bathtub), 6
ceiling fan w/light, 23
ceiling joist, 102
cellar, 102
cement fiber (asbestos) shingles, 102
ceramic tile, 36, 66
cesspool, 102
chain-link fence, 60
chain-link gate, 60
chair rail, 102
chime, door, 23
chimney, 103
chimney (metal), 42

chimney cap, 103
chimney stack, 103
clay tile, 56
clean sewer pipe, 50
clerestory window, 103
climate classification key, 82
closet bifold, 16
closet mirror, 16
closet pole, 61
closet siding, 16
closet, broom, 61
clothes dryer, 61
clothes washer, 61
collar beam, 103
column, 103
columns (wood), 32
combination windows, 71
common wall, 103
compactor, trash, 47
composition shingle, 103
computing areas, 79
   parallelogram, 79
   irregular polygons, 80
   properties of a circle, 81
   trapezium, 81
   trapezoid, 79
   triangle, 80
concrete, 32
concrete basement floor, 103

# INDEX

concrete paving, 60
concrete tile, 56
condominium, 103
conduit, 103
conduit (flexible), 23
connection, fixture, 50
construction, wood frame, 111
contractor, general, 107
control panel, fire alarm, 24
control, lawn sprinkler, 60
cooler, evaporative, 105
coping, 56
copper, 56
corner brace, 103
cornice molding, 103
countertop, 46
coved ceiling, 104, 109
cover, swimming pool, 61
crawl space, 104
cripple stud, 104
cripple, window, 111
cupola, 104
curbs (concrete), 60
curtain wall, 104

## D

damaged floor, sand & finish, 37

decks, wood, 61
dehumidifier, 42
depreciation, 91
   condition modifier table, 95
   typical life expectancy in years (table), 97
desk (built-in), 63
detached dwelling, 104
determining average story height, 78
device, photocell, 25
diagonal subflooring, 104
dining room, room lighting standards, 19
dishwasher, 46
dishwasher (built-in), 46
disposal, garbage, 46, 50
distribution box, septic, 61
diving board, swimming pool, 62
door chime, 23
door screen (complete), sliding, 14
door system, exterior, 13
door system, interior, 13
door trim, 16, 104
door, garage, 106
door, shower, 6
door, stain, 15, 17
doorbell/chime transformer, 23

# INDEX

doorjamb, 104
dormer, 104
double glazing, 104
double-hung window, 71, 104
downspout or leader strap, 56, 104
drain (floor), 50
drain (roof), 50
draperies, 63
drapery rod, 63
drapery track, 63
dryer, clothes, 63
drywall, 104
drywall finish system, 65
duct, air, 42
ducts, 105
dutch door, 14

## E

eave trough or gutter, 105
eaves, 105
elastomeric (single ply), 56
electric cable heating, 105
elevated slab, 105
elevation, 105
enclosure, bathtub, 6
enclosure, patio, 64
entrance canopy, 105

entrance platform, 105
entrance post, 105
entrance, room lighting standards, 19
entry door frame, 14
entry door hardware, 14
entry door trim, 14
entry door, metal standard, 14
entry door, wood custom, 15
evaporative cooler, 105
exchanger, air, 42
exhaust fan, 42, 43
expansion tank (hot water), 43
expansion tank insulation, 43
explanation of depreciation, effective age, 93
explanation of depreciation tables, 92
    extended life expectancy, 92
    remaining life, 92
exterior door system, 13
exterior fixture (decorative), 23
exterior fixture (plain), 23
exterior fixture (security), 23
exterior wall terminology
    frame system, 27
    masonry system, 28
external obsolescence, 92

# INDEX

## F

face brick, 105
fan (window), 43
fan, ceiling, 23
fascia, 56, 105
faucet, 6, 46
faucet, combination, 46, 50
faucet, double, 50
faucet, single, 50
felt paper, 56
fence, chain-link, 60
fence, redwood, 61
fence, split-rail, 62
fence, wood board, 62
fence, wood picket, 62
fence, wrought iron, 62
fenestration, 105
fiberglass, 56
fiberglass (corrugated), 56
filter, swimming pool, 62
filter, water, 51
finish floor, 105
finish hardware, 105
fire alarm control panel, 24
fire alarm station, 24
fireplace (masonry w/heatilator), 63
fireplace (masonry), 63
fireplace (prefab), 63
fireplace chimney (masonry), 63
fireplace hearth, 105
fireplace hearth (raised), 63
fireplace log lighter, 63
fixed windows, 71
fixture connection, 50
fixture rough-in, 50
flagpole, 60
flagstone/tile paving, 60
flashing, 56
floor area, 106
floor joists, 106
floor sheathing (boards), 36
floor sheathing (plywood), 36
floor sleepers, 36
flooring system, typical, 35
fluorescent fixture, 24
footing, 106
forced-air heating, 106
forced-air furnace, 39
formica, 106
foundation drain tile, 106
foundation wall, 106
fountain (decorative), 50
frame, door, 16
frame, entry door, 14

# INDEX

frame, window, 111
framing, stud, 33
freezer, 46
french door, 14, 16
french doors or windows, 106
fresco, 106
frieze, 106
functional obsolescence, 91
    curable functional obsolescence, 92
    incurable functional obsolescence, 92
furnace (forced air), 43
furring, 11, 32
furring strips, 106

## G

gable roof, 106
gable stud, 106
gambrel roof, 106
garage cornice, 106
garage door, 106
garage door (overhead sectional), 14
garage door (overhead single-leaf), 14
garage door opener, 14
garbage disposal, 46, 50

gate, chain-link, 60
gate, ornamental iron, 61
general contractor, 107
girder post, 107
glass block, 107
glass, insulating, 71
glass, opaque, 71
glass, plate, 71
glass, single, 72
glass, tempered, 72
glass, wired, 72
glazing, skylight, 72
glossary explanation, 99
grade line, 107
gravel stop, 56
gravity heating, 107
greenhouse, 63
greenhouse windows, 71
grille, air intake, 42
grilles, security, 72
grounding rod, 24
grout, 107
gutter, 56
gypsum board, 66
gypsum board (standard), 11
gypsum board (water resistant), 11, 66

# INDEX

## H

hallway, room lighting standards, 20

hardboard, 66, 107

hardboard boards, 32

hardboard panels, 32

hardware, 71

hardware, door, 16

hardware, entry door, 14

hardware, metal standard entry door, 14

header, 107

hearth, fireplace, 105

heat pump, 39, 107

heat, radiant ceiling, 44

heat, radiant floor, 44

heater (electric baseboard), 43

heater (hot water baseboard), 43

heater (tankless), hot-water, 50

heater wall, 43

heater, hot-water, 50

heater, swimming pool, 62

heating unit, sauna, 64

heating, baseboard, 100

height measurement technique, 77

hipped roof, 107

hollow-core wood door & hardware, 16

home improvement loan and mortgage payment tables, 88

home mortgages, types of, 85

hood, range, 46

hookup, appliance, 50

hose bib, 50

hot tub, 60

hot tub/spa, 63

hot water, 40

hot-water heating, 107

hot-water heater, 50

hot-water heater (tankless), 50

hot-water tank insulation, 50

humidifier, 43, 107

## I

incandescent fixture (decorative), 24

incandescent fixture (plain), 24

incandescent fixture (recessed), 24

insulating board, 66

insulating glass, 71

insulation, 11, 107

insulation (batt), 32, 57, 66, 100

insulation (blown-in), 32

# INDEX

insulation (rigid), 32, 57, 66
insulation requirements, 85
insulation, expansion tank, 43
insulation, hot-water tank, 50
intercom speaker, 24
intercom station, 24
intercom station w/radio, 24
interim money, cost of, 108
interior door system, 13
irregular polygons, 80

## J

jalousie, 71, 108
jalousie glass (only), 71

## K

keene's cement, 108
key, climate classification, 82
kitchen, room lighting standards, 20

## L

ladder, swimming pool, 62
landscaping, 60
lath, 66, 108
lawn sprinkler control, 60
lawn sprinkler head, 60

layout, typical kitchen, 45
leaching lines, septic, 61
light, window, 111
lighting, 61
lightning arrester, 25
linoleum, 36
lintel, 108
living & family, room lighting standards, 20
log lighter, fireplace, 63
lookout, 108

## M

mailbox (post type), 61
mailbox (wall type), 64
mansard roof, 108
mantle, 108
marble, 36
masonry block, 32
masonry construction, 108
masonry, repoint, 33, 68
measuring floor areas, 78
medicine cabinet, 6
mesh, 108
mesh, security, 72
metal, 57
metal lath, 108
metal standard entry door, 14
metal tile, 66

# INDEX

microwave oven, 46
millwork, 108
mirror, 6, 66
mirror tile, 66
mirror, closet, 16
modular construction, 108
molding base, 67
molding, ceiling, 67
molding, chair rail, 67
molding, corner, 67
molding, panel edge, 67
molding, cornice, 103
monolithic construction, 108
mortar, 108
mud sill, 109

## N

new floor, sand & finish, 37
newel post, 109

## O

oil tank supply line, 43
opaque glass, 71
ornamental iron gate, 61
outlet box, 25
outlet, telephone/TV, 26
oven, 46

oven, microwave, 46
overhead and profit, 109

## P

package unit, 44
pad, carpet, 36
paint, 11, 32, 67, 71
paint (textured), 11
paint door, 14, 16
panel, 109
panel door, raised-wood, 17
panelboard (circuit breakers), 25
paneling (plywood), 67
paneling (solid), 67
panels, hardboard, 32
panels, plywood, 12
paper, felt, 56
parallelogram, 79
parameter, 109
parapet wall, 109
parquet flooring, 109
patio enclosure, 64
pavers, brick, 36
paving, asphalt, 60
paving, brick, 60
paving, concrete, 60
paving, wood, 62

# INDEX

payment tables, home improvement loan and mortgage, 88

perimeter, 109

photocell device, 25

physical depreciation
   curable physical depreciation, 91
   incurable physical depreciation, 91

pier, 109

pilaster, 109

pilings, 109

pipe (hot water branch), 44

pipe (hot water main), 44

piping (cast iron), 50

piping (copper), 51

piping (galvanized steel), 51

piping (plastic), 51

plaster, 11, 67, 109

plaster & lath, 67

plaster repair, 67

plaster w/gypsum lath, 11

plaster w/metal lath, 12

plaster w/wood lath, 12

plastic panels (only), 12

plastic panels w/suspension grid, 12

plastic tile, 57, 67

plate glass, 71

platform, entrance, 105

plumbing fixtures, 109

plywood, 109

plywood (textured), 33

plywood panels, 12

pocket door, 16

pocket door & hardware, 16

pole, closet, 63

porch, 110

post, entrance, 105

precast concrete, 110

pressure tank, well water, 51

prime coat, 110

properties of a circle, 81

pump (circulating), 51

pump (sump), 51

pump, well, 51

# Q

quantity survey, 110

quarry tile, 36

# INDEX

## R

R-value, 110
radiant ceiling heat, 44
radiant floor heat, 44
radiant heating, 110
radiator, 44
radiator (fin tube), 44
rafters, 110
rail, chair, 102
railings, 64
raised-wood panel door, 17
ramps (concrete), 61
range (cooktop), 46
range hood, 46
range/microwave combination, 46
range/microwave/oven combination, 46
range/oven, 47
range/oven (built-in), 47
receptacle (108 volt), 25
receptacle (220 volt), 25
receptacle (exterior), 25
redwood fence, 61
refinish swimming pool, 61
refrigerator, 47
regrout tile, 67

regrout tile floor, 36
reinforcing steel, 110
repair existing plaster, 12
repair swimming pool cracks, 61
repair swimming pool tiles, 61
repair wood fence, 61
repair, plaster, 67
replaster swimming pool, 61
repoint masonry, 33, 68
requirements, insulation, 83
resilient, 37
resilient floor covering, 110
retaining wall, 61
ridge, 110
ridge board, 110
riser, 110
rod, drapery, 63
rod, grounding, 24
rod, shower, 6
roll roofing (composition), 57
roof boards, 110
roof styles, typical, 53
roof terminology, 54
roofing, 110
roofing materials, typical, 55
room lighting standards
    bathroom, 19
    bedroom, 19

# INDEX

dining room, 19
entrance, 19
hallway, 20
kitchen, 20
living & family, 20
site vegetation, 21
stairs, 21
study, 21
track, 21
room, sauna, 64
rough-in, 110
rough-in, fixture, 50
rubber tile, 37

## S

sandblasting, 33
sanding, 33
sash, 111
sauna heating unit, 64
sauna room, 64
screen door (complete), 14
screens, 72
seat, toilet, 6
security alarm base, 25
security alarm points, 25
security grilles, 72
security mesh, 72
septic distribution box, 61

septic leaching lines, 61
septic tank, 61, 111
service (1 phase), 25
service shut-off valve, 51
sewer pipe, clean, 50
shades, 64
shake, 111
shakes, wood, 34, 57
sheathing, 33, 57
sheathing, wall, 112
shelving, 64
shingles, composition, 57
shingles, miscellaneous, 33
shingles, wood, 34, 57
shower door, 6
shower rod, 6
shower stall, 6
showerhead, 6
shutter, 111
shutters (interior), 64
shutters (louver), 72
shutters (plain), 72
sidewalk, 111
siding strips, batten, 32
siding, aluminum, 32
siding, asbestos, 32
siding, closet, 16
siding, vinyl, 33
sill, 111

# INDEX

single glass, 72
single-hung windows, 72
sink, built-in, 6
sink, kitchen, 47
sink, laundry, 51
sink, wall mounted, 6
sink, wet bar, 51
site vegetation, lighting standards, 21
skylight, 57, 111
skylight glazing, 72
slab, elevated, 105
slate, 37
slate tile, 57
sleepers, floor, 36
sliding door (aluminum /vinyl-complete), 14
sliding door, wood, 15
sliding door screen (complete), 14
sliding windows, 72
slope, 111
smoke detector, 111
soffit, 57
softener, water, 51
solar heating systems, 41
spa, 62
speaker, intercom, 24
spiral stairs, 64

split-rail fence, 62
spotlight (decorative), 25
sprinkler head, lawn, 60
stain, 12, 33, 68
stain door, 15, 17
stair riser, 111
stair stringer, 111
stairs, room lighting standards, 21
stairs, spiral, 64
stall, shower, 6
station w/radio, intercom, 24
station, intercom, 24
steps, 62
stock entry door, wood custom, 15
stock entry door, wood standard, 15
stone veneer, imitation, 33, 68
stone veneer, natural, 33
stone wall, 62
stone wall patterns, 30
stop, gravel, 56
storm door, 111
storm door, aluminum, 14
storm door, wood, 15
storm window, 72, 111
stove (wood/coal burning), 64
stucco, 111
stucco (on framing), 33

# INDEX

stucco (on masonry), 33
stud, 111
stud framing, 33
stud, corner, 103
studs, 68
study, room lighting standards, 21
subflooring, diagonal, 104
subpanel, distribution, 23
sump pump, 112
supply line, oil tank, 43
suspension grid, 12
swimming pool, 62
swimming pool cover, 62
swimming pool cracks, repair, 61
swimming pool diving board, 62
swimming pool filter, 62
swimming pool heater, 62
swimming pool ladder, 62
swimming pool tiles, repair, 61
swimming pool, refinish, 61
swimming pool, replaster, 61
switch, 3-way, 25
switch, dimmer, 25
switch, exterior, 25
switch, wall, 26
system, heat pump, 43

## T

tank, oil, 43
tar paper strip, 112
telephone, 26
telephone base station, 26
telephone/tv outlet, 26
tempered glass, 72
terminology, roof, 54
termite shield, 112
terra cotta, 112
terrazzo, 112
thermostat, 26, 44, 112
thermostat (programmable), 26, 44
threshold, 15, 17
tie, 112
tile floor, regrout, 36
tile paving, flagstone, 60
tile, asphalt, 36
tile, ceramic, 36, 66
tile, clay, 56
tile, concrete, 56
tile, metal, 66
tile, mirror, 66
tile, plastic, 57, 67
tile, quarry, 36
tile, regrout, 67
tile, rubber, 37

# INDEX

tile, slate, 57
tile, vinyl, 37
tile, wood parquet, 37
timer, 26
toilet, floor-mounted, 6
toilet, wall-mounted, 7
toilet seat, 6
tongue and groove (T&G), 112
track lighting, 26
track, drapery, 63
track, room lighting standards, 21
transformer, doorbell/chime, 23
trapezium, 81
trapezoid, 79
trash compactor, 47
triangle, 80
trim, 68, 72
trim, wood/metal, 33
trim, door, 16
trim, entry door, 14
tank, septic, 61
tub, hot, 60
tubing (copper), 51
types of home mortgages, 85
typical bathroom layout, 5
typical building lives, 94
typical ceiling system, suspended acoustical, 10
typical ceiling system finishes, drywall, 9
typical exterior wall systems, brick or stone veneer, 31
typical flooring system, 35
typical garage styles, 75
   attached garages, 75
   basement garages, 75
   built-in garages, 75
   carports, 76
   detached garages, 75
typical house styles, 73
   mobil/manufactured housing, 74
   multiples, 74
   one and one half story, 73
   one story, 73
   split level, 74
   town houses and duplexes, 74
   two and one half story, 73
   two story, 73
   two story bi-level, 73
   urban row houses, 75
typical kitchen layout, 45
typical plumbing layout and pipe sizes, 49
typical roof styles, 53
typical roofing materials, 55

# INDEX

typical windbreak layout, 59
typical wiring layout, 22

## U

U factor, 112
underlayment (hardboard), 37
unit, package, 44

## V

valve, service shut-off, 51
vanity, metal, 7
vanity, wood, 7
vapor barrier, 112
veneer, 112
veneer, brick, 32, 66
vent (dryer), 44
vent stack, 44
ventilator (attic), 44
vinyl composition (asbestos) tile, 112
vinyl siding, 33
vinyl tile, 37

## W

wainscoting, 68, 112
wall cabinet, 47
wall furnace, 41

wall sheathing, 112
wall, areaway, 99
wall, retaining, 61
wall, stone, 62
wallpaper, 12, 68
wallpaper (grass cloth), 68
wallpaper (mural), 68
walls, block, 60
walls, brick, 60
wardrobe, 64
washer, clothes, 63
water filter, 51
water softener, 51
waterproofing, 113
waterproofing, building paper, 33
waterproofing, cement parging, 33
waterproofing, hot-mopped, 34
waterproofing, plastic sheeting, 34
weatherstripping, 113
well (drill & case), 51
well pump, 51
well water pressure tank, 51
window casing, 113
window cripple, 113
window frame, 113
window light, 113
window terminology, 70

# INDEX

window unit types, 69

windows, awning, 71

windows, casement, 71

windows, combination, 71

windows, double-hung, 71

windows, fixed, 71

windows, greenhouse, 71

windows, single-hung, 72

windows, sliding, 72

windows, storm, 72

wired glass, 72

wiring (108 volt), 26

wiring (220 volt), 26

wiring coaxial (TV/radio), 26

wiring layout, typical, 22

wiring low voltage (doorbell/thermostat), 26

wood (hardwood), 37

wood (softwood), 37

wood beams, 12

wood board fence, 62

wood custom entry door, 15

wood custom stock entry door, 15

wood decks, 62

wood door, flush hollow-core, 16

wood door, flush solid-core, 16

wood door & hardware, flush hollow-core, 16

wood fence, repair, 61

wood frame construction, 113

wood parquet tile, 37

wood paving, 62

wood picket fence, 62

wood plank, 12

wood shakes, 34, 57

wood shingles, 34, 57

wood siding (bevel), 34

wood siding (clapboard), 34

wood standard stock entry door, 15

wood storm door, 15

workbench, 64

wrought iron fence, 62

# Home Repair & Remodel Cost Guide

## THE MOST UP-TO-DATE, COMPLETE COST GUIDE AVAILABLE TODAY!

For: Realtors, Appraisers, Homeowners, Contractors, Etc...

With Marshall & Swift's *Home Repair & Remodel Cost Guide*, you can quickly discover just how much it will cost to perform almost any home repair or remodel job. This easy-to-use, comprehensive guide is a powerful negotiating tool, a helpful budgeting resource, as well as a handy estimating guide.

- **Discover why we say, "Open The Book and Close The Sale!"**
- **Sell more homes and justify asking prices to buyers**
- **Provide a quick "guesstimate" on repairing a fixer-upper**
- **Provide homeowners with accurate remodeling costs**
- **Perfect for "Cost to Cure" estimates or for Comparative Market Analysis (CMAs)**

In addition to costs, definitions and illustrations, this guide features types of mortgages, home improvement loan payment tables, and depreciation tables. With basic terminology explained, detailed drawings, component prices, and directions for many repair and remodel jobs, the *Home Repair and Remodel Cost Guide* is a must-have reference tool for everyone.

Call Marshall & Swift now!
Order by calling toll-free:
**(800) 544-2678**

**Now Only $49.95!**

The most comprehensive and up-to-date information is now available at your fingertips.

**MARSHALL & SWIFT**
THE BUILDING COST PEOPLE

# Tenant Improvement Cost Book... for Office

## PROVIDING PRACTICAL AND CONVENIENT OFFICE INTERIOR COSTS!

For: Appraisers, Architects, Contractors, Interior Designers, Building Managers, Etc...

The *Tenant Improvement Cost Book ... for Offices* takes you through a "checklist" of what's standa office interior construction so you won't overlook anything! You can cost out the office space on pc before you even hammer in the first nail! This helps architects, contractors and interior designers p and design projects with a full view of what it will cost to build out the interior office space.

- **Commercial real estate professionals can help guide clients' decisions**
- **Easy-to-understand diagrams show how to make the most of your available space**
- **Helpful cost-cutting hints for times when you are renovating existing space**
- **Learn the tricks of effective space planning**
- **Gain a better understanding of how much space is required, based on your specific nee**
- **Find more than 7,000 costs for typical office components like carpeting, framing & mor**

Included is a complete glossary of construction and office leasing terms with construction components broken down by building system categories like Finishes, Wood & Plastics and Electrical. You can eas work through your cost estimate component by component! And when you're filling out the workshee you have the flexibility of choosing the costs of your components using different quality levels.

Get your complete guide to estimating interior office improvements now!

## Call Marshall & Swift now!
### Order by calling toll-free:
# (800) 544-2678

**MARSHALL & SWIFT**
THE BUILDING COST PEOPLE

*Tenant Improvement Cost Book... for Offices* 199

***Now only $95.95!***

"The Marshall & Tenant Improven Cost Book ... for Offices provides practical, conver office interior cc and so much mc

# The Residential Cost Handbook!

## THE ONE ANSWER TO ALL YOUR RESIDENTIAL COST QUESTIONS...

For: Appraisers, Assessors, Lenders, Builders, Contractors, Realtors, Etc.

Marshall & Swift's *Residential Cost Handbook* contains costs from simple to complex (and even unusual) residences. Its comprehensive yet easy-to-use format makes it the one easy answer to all your residential cost questions.

**Enjoy the flexibility of choosing the detail level of three different cost methods!**
Square Foot, Segregated and Comparative

**Enjoy the confidence of current, accurate estimates, constantly updated and sent to you quarterly.**
Photos, Charts, Cost Modifiers, Illustrations and Glossary

**Enjoy the ease of using one, single source!**
Single Family, Town Houses, Log Cabins, Mfd. Homes, Low-rise Apts. and Cottages

The *Residential Cost Handbook*'s cost approach gives you the cost of a home even when there are no comparables! Whatever the challenge, this is the answer to all your residential cost questions.

**Call Marshall & Swift now!**
Order by calling toll-free:
**(800) 544-2678**

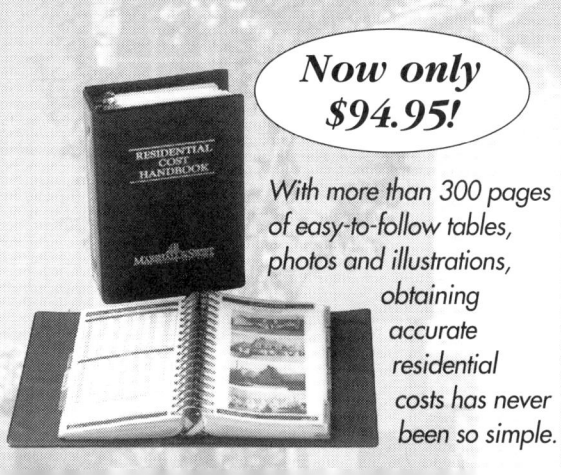

*Now only $94.95!*

With more than 300 pages of easy-to-follow tables, photos and illustrations, obtaining accurate residential costs has never been so simple.

**MARSHALL & SWIFT**
THE BUILDING COST PEOPLE

# The Dodge Repair & Remodel Cost Book

## FIND COMPONENT COSTS QUICKLY AND EASILY...

For: Contractors, Architects, Adjusters, Realtors, Etc.

Quickly find labor, material and complete component costs for residential and light commercial construction e mating. Save yourself the trouble of calling and running around after obscure component costs. They're all he this easy-to-use, comprehensive guide to over 10,000 component costs.

**Comprehensive –**
- 300 pages containing over 10,000 component costs

**Current –**
- Depend on accurate labor and material costs, continually researched and updated annually

**Convenient –**
- Organized by the Construction Specifications Institute (CSI) Masterformat™
- Easily find labor output per day and man-hours
- Confidently rely on construction costs adjusted geographically with local multipliers.

The next time you need component costs for residential or light commercial renovation, reach for the one compreher current, and convenient source thousands of professionals depend on—*The Dodge Repair & Remodel Cost Book.*

**Call Marshall & Swift now!**
Order by calling toll-free:
## (800) 544-2678

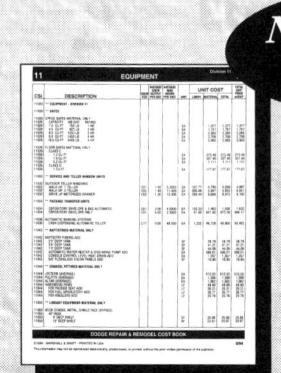

**MARSHALL & SWIFT**
THE BUILDING COST PEOPLE

*Now only $59.95!*

A typical cost page filled wi labor & mate costs.

# Dodge Unit Cost Book

## THE PERFECT REFERENCE FOR A QUICK COST LOOK UP!

**For: Contractors, Assessors, Appraisers, Architects, Etc...**

The *Dodge Unit Cost Book* is an excellent professional tool for anyone estimating new construction projects, establishing preliminary budgets or just spot checking estimates. Labor, material, equipment, total costs with overhead and profit are at your fingertips!

**This publication contains:**
- **Over 15,000 units-of-construction costs**
- **Labor, material and equipment costs**
- **Overhead, profit, crew size, productivity rates and prices for hard-to-find-items**

The book enables you to make adjustments to customize your costs. It's easy to calculate for local labor and material costs by using the handy local multipliers for more than 825 geographic regions throughout the United States and Canada. All components are conveniently arranged according to the Construction Specification Institute's (CSI) Masterformat™. All costs included in the *Dodge Unit Cost Book* are based on the widely respected Marshall & Swift national average.

Marshall & Swift has one of the largest database of comprehensive building costs. The research team tracks actual labor and material costs while monitoring the factors and technologies that influence construction costs. Labor rates are gathered for 22 different trades in each of the researched cities. The local multipliers are based on prevailing wages for trades within each region. Material costs are determined by contacting building product manufacturers, dealers, supply houses, distributors and contractors.

**Call Marshall & Swift now!**
**Order by calling toll-free:**
## (800) 544-2678

**MARSHALL & SWIFT**
THE BUILDING COST PEOPLE

*Now only $59.95!*

*The* Dodge Unit Cost Book *is an indispensable resource for quickly and easily building estimates, establishing budgets or spot checking costs.*

# Dodge Metric Unit Cost Book

## GET GOVERNMENT BIDS... USE THE METRI[C] SYSTEM TO CALCULATE COST ESTIMATES

For: Contractors, Assessors, Appraisers, Architects, Etc...

The *Dodge Metric Unit Cost Book* is the most comprehensive cost guide for residential and ligh[t] commercial construction in the industry today!

**This publication contains:**
- **Over 15,000 units-of-construction costs**
- **Labor, material and equipment costs**
- **Overhead, profit, crew size, productivity rates and prices for hard-to-find-item[s]**

The book enables you to make adjustments to customize your costs. It's easy to calculate for local labor and material costs by using the handy local multipliers for more than 825 geographic region[s] throughout the United States and Canada. All components are conveniently arranged according t[o] the Construction Specification Institute's (CSI) Masterformat™. All costs included in the *Dodge Metri[c] Unit Cost Book* are based on the widely respected Marshall & Swift national average.

Marshall & Swift has one of the largest database of comprehensive building costs. The research te[am] tracks actual labor and material costs while monitoring the factors and technologies that influence construction costs. Labor rates are gathered for 22 different trades in each of the researched cities The local multipliers are based on prevailing wages for trades within each region. Material costs are determined by contacting building product manufacturers, dealers, supply houses, distributors and contractors.

**Call Marshall & Swift now!**
**Order by calling toll-free:**
# (800) 544-2678

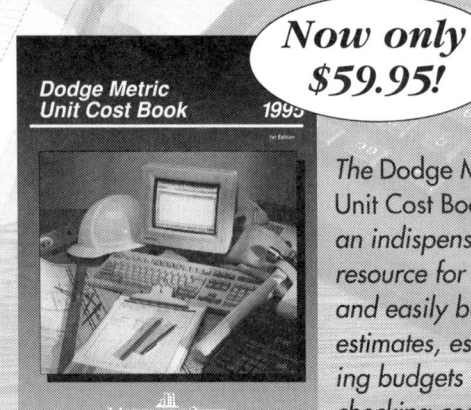

*Now only $59.95!*

*The* Dodge Metr[ic] Unit Cost Book i[s] an indispensabl[e] resource for qui[ck] and easily build[ing] estimates, estab[lish]ing budgets or s[imply] checking costs.

## MARSHALL & SWIFT
THE BUILDING COST PEOPLE

# The Marshall Valuation Service

## YOUR SINGLE SOURCE FOR COMMERCIAL BUILDING COSTS

For: Architects, Developers, Lenders, Real Estate Professionals, Appraisers, Tax Assessors, Etc...

*Marshall Valuation Service* is the most authoritative and comprehensive guide to commercial and industrial building costs. Discover for yourself just how easy it is to obtain costs for even the most complex buildings.

**Over 30,000 Component Costs**

**Costs for over 150 types of commercial and industrial buildings:**
Farm Sheds, High-Rises, Health Clubs, Restaurants, Retail Stores, Warehouses, Etc...

**A wide variety of Supplemental Costs:**
Car Washes, Greenhouses, Storage Tanks, Etc...

**Plus costs for:**
Landscaping, Recreation Facilities, Subdivisions and Yard Improvements

**Local Multipliers adjust for over 800 geographical areas in:**
U.S.A., Guam, Canada and Puerto Rico

**Use the Estimating Method that suits your need:**
*Calculator* (square foot) *Method* – for simple buildings or quick estimates.
*Segregated Method* – reconstruct complicated structures from the foundation up.
*Comparative Cost Multiplier Method* – trend known costs forward or back in time.

Enjoy the confidence of using the one single comprehensive source thousands of professionals have learned to depend on!

**Call Marshall & Swift now!**
Order by calling toll-free:
**(800) 544-2678**

*Now only $169.95!*

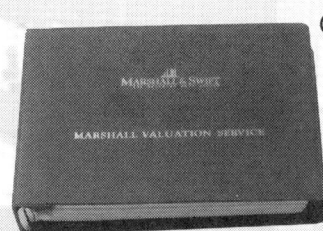

*Quickly find the costs you need in this 400-page, monthly updated guide to commercial building costs.*

**MARSHALL & SWIFT**
THE BUILDING COST PEOPLE

# The Commercial Estimator

## COMMERCIAL AND INDUSTRIAL COST ESTIMATING HAS NEVER BEEN EASIER!

For: Appraisers, Assessors, Architects and Realtors

For more than 60 years, Marshall & Swift has been collecting, refining and creating building cost data products. Now, let our comprehensive database provide you with the most up-to-date information in seconds.

- **Eliminate manual adjustments**
- **Customize estimates to your exact specifications**
- **Save time**
- **Reduce your margin of error**
- **Automatically generate professional reports**
- **Use accurate and reliable data, updated quarterly**

Take the opportunity to experience firsthand the Speed, Precision and Power of this business enhancement tool. We invite you to purchase *Commercial Estimator 5.0* at the discounted price of $499.00. You save $160.00 off the regular price!

Call Marshall & Swift now!
Order by calling toll-free:
**(800) 544-2678**

*Now only $499.00!*

The most comprehe... and up-to informatio now avail. at your fingertips.

**MARSHALL & SWIFT**
THE BUILDING COST PEOPLE

# The Residential Estimator Software

## AUTOMATE YOUR RESIDENTIAL ESTIMATES

**For: Appraisers, Assessors, Lenders, Builders, Contractors, Realtors, Etc.**

Now, you can use the speed and accuracy of digital data processing to estimate costs for all your residential valuations. Enjoy the power of Marshall & Swift's comprehensive database, constantly updated and sent to you quarterly on disk! With a few simple keystrokes you can have accurate, professional cost reports in your hand, ready for your client.

- **Uses the Square Foot Method**
- **Automatically adjusts for climate and location**
- **Adjusts for physical depreciation**
- **Calculates all the math automatically**
- **Comes in 3 versions: Assessor, Insurance or General/Appraiser**
- **Three professional reports (from a single entry): Standard, Summary or Detailed**
- **Two disk sizes (3-1/2" or 5-1/4")**

For just about $1.00 a day, and with a complete satisfaction guarantee, you have nothing to lose and much to gain. Call Marshall & Swift today! Order your copy of *Residential Estimator* toll-free.

## Call Marshall & Swift now!
### Order by calling toll-free:
# (800) 544-2678

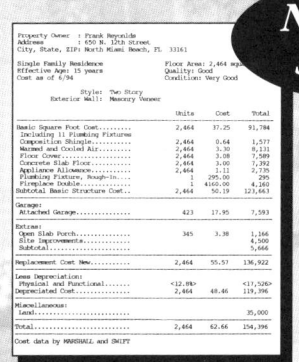

**Now only $379.00!**

*A sample Detailed Report from the Residential Estimator. Just enter the basic information and you'll be turning out professional reports.*

**MARSHALL & SWIFT**
THE BUILDING COST PEOPLE

# The Repair & Remodel Estimator

## AUTOMATE YOUR REPAIR AND REMODEL ESTIMATES

For: Realtors, Home Owners, Contractors, Appraisers, Etc.

In just seconds and with only a few simple keystrokes you can find the costs for renovating a dining room, or updating a kitchen. With Marshall & Swift's *Repair & Remodel Estimator* you h all the information you need to easily estimate any residential and light commercial constructio

**Saves time and energy:**
- Simply enter the basic search parameters
- No forms needed
- Uses over 1,200 Component Costs

**Ensures accuracy:**
- Does all the look-ups for you
- Performs all the necessary calculations
- Choose between time and material vs. unit cost

**Promotes professionalism:**
- Automatically generates professional reports in minutes: (Location Summary, Component Report, Work Summary, Detailed Report, or Bid Report)

Save yourself the aggravation of hand-generated repair estimates. Use the power of Marshall & Swift's PC Repair and Remodel Software to automate your estimates.

## Call Marshall & Swift now!
### Order by calling toll-free:
# (800) 544-2678

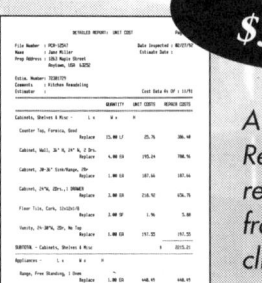

**Now only $349.00!**

A sample Detailed Report. Just one of reports you can ch from to impress yo clients or your bos:

## MARSHALL & SWIFT
THE BUILDING COST PEOPLE